MONOGRAPHS
ON APPLIED PROBABILITY AND STATISTICS

*General Editors:* M. S. BARTLETT F.R.S.
and D. R. COX F.R.S.

# SEQUENTIAL METHODS IN STATISTICS

# Sequential Methods in Statistics

G. BARRIE WETHERILL

*Professor of Statistics, University of Kent at Canterbury*

LONDON

CHAPMAN AND HALL

A Halsted Press book
John Wiley & Sons, Inc., New York

First published 1966
by Methuen & Co. Ltd
Second edition 1975
published by Chapman and Hall Ltd
11 New Fetter Lane, London EC4P 4EE
First issued as a paperback 1979



Printed in Great Britain at the
University Press, Cambridge

ISBN 0 412 21810 0



Distributed in the U.S.A. by Halsted Press,
a Division of John Wiley & Sons, Inc., New York

# Contents

# Preface

This book began some years ago when I was asked to lecture on 'Sequential Analysis' at the Institute of Engineering Production, University of Birmingham. A first draft of Chapters 1–4 and 6–8 appeared in 1963, during leave of absence from Birkbeck College, while I was with the Statistics Research Department of Bell Telephone Laboratories, Murray Hill, N.J. I am grateful for the generous co-operation of Bell Laboratories, in enabling me to continue this work during my year-long stay with them.

Sequential analysis has been heavily dominated by the sequential probability ratio test, and the time seems ripe to take stock of what has been achieved. In this book particular emphasis has been placed on methods which are of importance for making practical applications. Except for the last two chapters, I have attempted to include references to as many published applications as I could find, although I cannot claim to have a complete list. I present and discuss the logical basis of the methods, but I do not include proofs of theorems unless they are needed for the discussion of some extension or generalization. Problems are given to each chapter, and many of them do not have simple answers; some of them are problems for further research. In some cases part of the problem is to give a precise formulation. To refer to these problems, such as to problem 4 of Chapter 6, I use the notation P6.4.

I have tried to make the book understandable to applied statisticians as well as to students and research workers in theoretical statistics.

I am indebted to Professor D. R. Cox for encouragement to write the book and for much helpful criticism and advice. I am also greatly indebted to the general editor, Professor M. S. Bartlett, for many useful comments. Others who have discussed parts of the book, or with whom I have discussed material relating to it, are Professors

Anscombe, Armitage, Lindley, Dr. R. N. Curnow and Dr E. A. Gehan. Chapter 7 is based on work I did as a postgraduate student under the guidance of Professor G. A. Barnard. Errors are, of course, my responsibility.

CHAPTER 1

# Introduction

## 1.1. Sequential Experiments

Sequential experimentation is an area of statistics which is both of practical importance and also of great theoretical interest. In this volume we shall attempt to summarize the main theoretical results, and give some general discussion and criticism of the methods and formulations, in terms of practical applications.

Consider the following examples.

*Ex. 1.1* Box (1954, p. 45) has described experiments concerned with maximizing the yield of a chemical process. The process worked in two stages, and three factors could be varied at the first stage (temperature, time of reaction and concentration), and two at the second (temperature and time of reaction). A preliminary experiment was run using all five factors, and based on the results obtained, further experiments were run until the optimum working conditions were determined. For this example several equivalent sets of conditions were found for which the yield was equally high (approximately). Box and Wilson (1951), Box (1954) and Box and Hunter (1957), have considered in detail how such series of observations should be planned.

*Ex. 1.2.* Snoke (1956) has described experiments for studying the strength of wood preservatives. Some $\frac{3}{4}$-in. cubes of wood were sawn from selected wood, and the cubes carefully marked, dried and weighed, and then separated into groups representing different densities. The blocks were impregnated with a wood preservative, and great care was taken to see that the blocks were fully treated, but that all surplus preservative was wiped off before reweighing. The amount of preservative retained in the blocks was measured on a scale called the retention level. After further preparation, weathering, etc., the blocks were exposed to a culture for a 90-day period, under controlled temperature, humidity, etc. After the test period the blocks were lightly brushed free of growth, and reweighed. The experimental results show that percentage weight loss decreases roughly exponentially with retention level.

In Ex. 1.2 the experiment is lengthy, and takes very careful preparation, so that it must be planned in detail before any results are obtained; that is, the retentions desired, the cultures used, the number of blocks, etc., are all fixed in advance. Ex. 1.1 shows up in sharp contrast to this, for both the number of observations made and, after the first stage, the levels of the factors used, depend on the results obtained. For the particular case referred to (Ex. 1.1), the first stage yielded sufficient information to determine approximately the conditions giving an optimum yield, but further trials were necessary to obtain more precise knowledge. Usually the first stage will only yield enough information to indicate the directions in which factor levels must be altered to increase the yield, and several stages will be necessary to determine optimum conditions.

Ex. 1.1 illustrates the main features of what we shall call a sequential experiment. A sequential experiment may be defined as one in which the course of the experiment depends in some way upon the results obtained. This definition is very vague, but this need not trouble us; in the widest sense, the whole of life is sequential, for our future actions are conditioned to some extent by our past experience. Indeed, the scientific method is by its very nature sequential; for example, whether a suggested wood preservative is thought worthy of further study depends on the results of the first experiments on it. In this volume we restrict attention to experiments in which the sequential nature is formalized.

The sequential nature of Ex. 1.1 is exhibited in two ways. Firstly, there is the sequential choice of the experiment to be performed, in that the factor levels used in the second and succeeding stages depend on earlier results. Secondly, a rule for termination of the experiment has to be formulated; this rule should allow trials to continue until there is evidence that a combination of factor levels giving near maximum yield has been estimated. These two aspects of Ex. 1.1, although strongly interconnected, are separate problems. The first problem deals with strategies of how to use observations as they become available, so as to find quickly the (approximate) position of maximum yield, while the second aspect deals with how precisely the position of maximum yield is to be estimated before terminating the experiment.

Sequential analysis is usually delimited by experiments in which stopping rule considerations alone are involved, and Chapters 2 to 6 inclusive are devoted entirely to this. When it is only the number of observations in an experiment which is sequentially dependent, the theory tends to be rather simpler and of more general applicability than when levels of trials are also sequentially dependent, and most of the theoretical advances have been made in this area. Two important applications of sequential experimentation in which (usually) only the number of observations is sequentially dependent are sequential medical trials and sampling inspection.

Armitage (1960) has published an account of some of the considerations involved in sequential medical trials, and describes a number of applications in detail. All these trials are direct comparisons between two alternative treatments, and there is no question of any 'level' of a treatment being involved. Either both treatments are given in random order to each individual, or else individuals are paired, and the two treatments are randomly allocated to the individuals in each pair. Briefly, Armitage argues that ethical considerations demand that a trial be stopped as soon as there is clear evidence that one of the treatments is to be preferred, and this results in a sequential trial. For various reasons the theory and methods developed by Wald (see Chapters 2 and 3) were felt to be inappropriate, and Armitage and Bross developed other, more suitable, methods (see Chapter 6 for some details).

Suppose now that we wish to compare several drugs or treatments (as in sequential screening of cancer drugs, see § 12.2), then it should clearly be possible to drop some drugs out of the trials at an early stage, if the results from these compare unfavourably with others. Thus we are led to experimental designs in which both the number and 'level' of trials are sequentially dependent, the 'levels' in this case corresponding to different treatments. The determination of schemes setting out the precise conditions under which some drugs would be dropped and others considered further is an interesting theoretical problem to which there is as yet no completely satisfactory solution; see § 12.3.

The most rudimentary form of sequential procedure is the double sampling plan. This is simply a plan in which a sample of a fixed size is taken, and if there is sufficient evidence for the purpose for which the experiment was designed the experiment is terminated. If the evidence is not sufficient, a second sample of a fixed size is taken and the experiment terminated without further trial. A double sampling plan for sampling inspection was proposed by Dodge and Romig (1929). Since World War II a number of sampling inspection tables have been produced, and most of these have included both double sampling and fully sequential plans, see Hill (1962) and references. It would appear that sampling inspection has been the practical field from which the impetus came for the most important theoretical developments in sequential analysis. In the next section we shall consider sampling inspection in some detail, and use this application to illustrate some properties of sequential schemes, and also to introduce some important concepts we use later in the book.

## 1.2. Sampling Inspection

Suppose that batches of items are presented for acceptance inspection, and that each batch contains a large number of items which can be classified as effective or defective. For simplicity we shall assume that each batch is to be sentenced independently, and that the sentence is one of two decisions; to accept or reject; to sell at the usual price or sell at a reduced price, etc. The most commonly used plan is a single sampling plan, in which a fixed number $N$ items are inspected in each batch, and the batch is rejected if $c$ or more defectives are found. Otherwise the batch is accepted.

The results of inspection in the set-up just described can be presented graphically in the lattice diagram. The number of defective items obtained (denoted $r$) is plotted vertically, and the number of effective items obtained (denoted $s$) is plotted horizontally. (Alternatively, $r$ could be plotted vertically, and the total number of items sampled $(r+s) = N$ plotted horizontally.) For the single sampling plan just described, the sampling results must be one of the points on the line $r+s = N$, since exactly $N$ items are inspected in every batch, and further, all points on the line for which $r \geqslant c$ are rejection points, and all points on the line for which $r \leqslant c-1$ are acceptance points. If the $N$ items are inspected successively the course of inspection traces out a path starting at the origin and ending at any point on the line $r+s = N$. In the diagram we have fixed $N = 20$, $c = 3$, and one such sample path is given, ending at a rejection point.

Now the single sampling plan given above is inefficient in the following sense – that if the first five or ten items inspected are all defectives, we shall continue to inspect items from the batch until we have inspected $N$ in all, even if more than $c$ defectives have been found. A very simple change eliminates this waste of effort, and makes this fixed sample size (non-sequential) plan into a primitive form of sequential plan.

Consider the following sampling plan. Proceed inspecting items until: either $r \geqslant c$ when we stop inspecting and reject the batch,

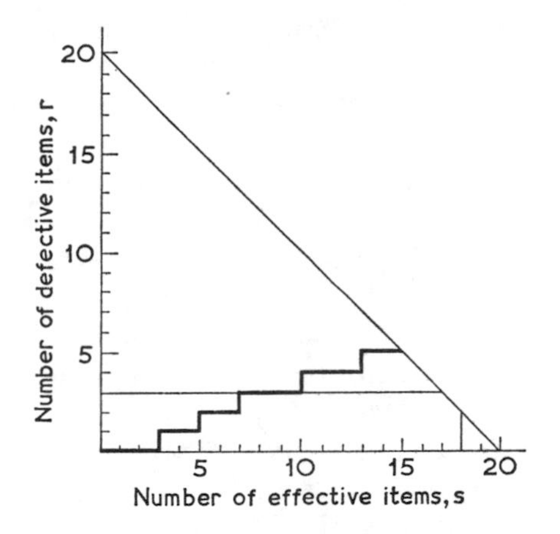


FIG. 1.1. The lattice diagram; $N = 20$, $c = 3$.

or $s \geqslant N-c+1$ when we stop inspecting and accept the batch. This is equivalent to having two boundaries drawn on the lattice diagram, a horizontal line at $r = c$, and a vertical line at $s = N-c+1$, and inspection stops as soon as either boundary is reached. This sampling plan is *sequential*, since we must stop after each item inspected and see if either of the conditions given above is reached, and if not we inspect another item. The sample size required is not fixed but variable; it may be as small as $c$ (if all $c$ items were defectives), or it may be as large as $N$. This sampling plan is called curtailed inspection.

It is easy to see that for all possible sample paths the sentences made by curtailed inspection and the equivalent single sampling plan are the same. In adopting this simple form of sequential sampling plan therefore we may save a substantial amount of inspection, while losing nothing.

Some advantages and disadvantages of sequential sampling plans are immediately apparent. (Although we shall base this discussion on the above example, most of the points mentioned here have general validity.)

Firstly, the sequential sampling plan is more complicated to operate than the fixed sample size plan, and hence inspection is slower, since inspectors have to stop after each item selected and check whether various conditions are fulfilled. The decision boundaries in the lattice diagram (Fig. 1.1) can in fact be modified to improve the average sample size properties of curtailed inspection, without causing much alteration to other properties of the plan. Such modifications will lead to plans still more complicated to operate. In practice, the requirements to be made of a sampling plan are often difficult to formulate precisely, and it may be preferable to choose a simple sequential plan rather than one which satisfies some neat optimality criterion (which will rarely be exactly appropriate to a given practical situation). One method of simplifying sequential plans is to sample items in groups, rather than singly; this both simplifies the plan and reduces the amount of checking to be done.

Secondly, the demand on labour is variable, and this itself may produce administrative difficulties. This variability can be reduced by inspecting in groups at a time.

Thirdly, suppose we operate a curtailed sampling plan for $N = 20$, $c = 3$. If the first three items chosen are defective, inspection stops, and the batch is rejected. Although we have enough information to make the correct sentence on the batch, curtailed inspection has provided us with only a very poor estimate of the percentage of defectives in the batch. We cannot, however, criticize the curtailed inspection plan for failing to satisfy requirements that it was not designed to meet. If our requirements can be precisely formulated, sequential plans can – in principle – be developed to meet them, but the formulation of the requirements should be carefully thought over. It is often the case that sequential plans to satisfy the particular requirements of a given situation have not been evaluated.

Against all these objections we have the very real advantage that while the curtailed sampling plan and the single sampling plan come to exactly the same sentence for any given sample, the average amount of inspection may be very much reduced by operating the sequential plan. It will be helpful at this stage to work out in detail some of the properties of curtailed inspection, so that we shall be able to see more clearly where its advantages lie.

## 1.3. Properties of the Curtailed Inspection Plan

Suppose the batches being inspected are very large in comparison with the sample size, then if a batch has a proportion $\theta$ of defectives

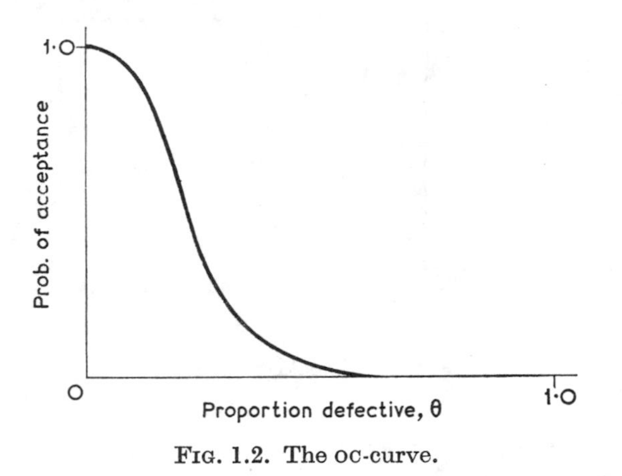


FIG. 1.2. The OC-curve.

the probability that it will be accepted is

$$\Pr\{\text{accept}|\theta\} = \Pr\{r \leqslant c-1|\theta\} = \sum_{r=0}^{c-1} \binom{N}{r} \theta^r (1-\theta)^{N-r}$$

$$= S(N, \theta, c-1), \tag{1.1}$$

say. This probability is the same either for curtailed inspection, or for the equivalent single sampling plan. The function (1.1) is called the *operating characteristic*, or, when plotted, the OC-*curve*, see Fig. 1.2. It is clear that for 'reasonable' plans, the OC-function is unity when $\theta = 0$, and zero when $\theta = 1$. An ideal OC-curve can sometimes be thought of as given in Fig. 1.3a, where

$$\Pr\{\text{accept}|\theta\} = \begin{cases} 1 & (\theta < \theta') \\ 0 & (\theta > \theta') \end{cases},$$

but it would be necessary to inspect the whole of each batch to achieve this OC-curve (unless $\theta' = 0$ or 1). Another way of formulating requirements on the OC-curve is to require

$$\Pr\{\text{accept}|\theta\} = \begin{cases} 1 & (\theta < \theta_0) \\ 0 & (\theta > \theta_1) \end{cases},$$

where the interval $(\theta_0, \theta_1)$ is an indifference region, and it is of small importance which decision is taken, see Fig. 1.3b. Even this relaxed

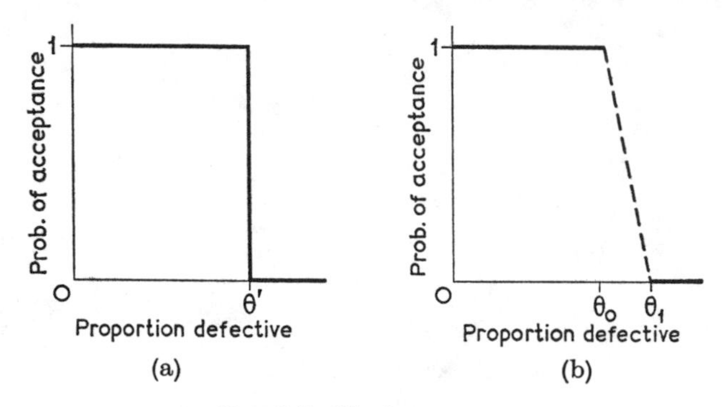


FIG. 1.3. Ideal OC-curves.

form of requirement cannot be attained exactly without 100% inspection, but it can be modified as follows,

$$\Pr\{\text{accept}|\theta\} = \begin{cases} 1-\alpha & (\theta = \theta_0) \\ \beta & (\theta = \theta_1) \end{cases}. \tag{1.2}$$

The parameters $N$ and $c$ of the curtailed inspection plan could be chosen to satisfy (1.2) exactly, except for the discreteness of the distribution. In this form the requirements are equivalent to a size $\alpha$ and power requirement $\beta$ in the Neyman–Pearson theory of testing hypotheses.

A second property of curtailed inspection which we consider is the sample size distribution. If a batch is rejected, the sample size ranges from $c$ to $N$, and

$$\Pr(n = c) = \theta^c$$

$$\Pr(n = c+r) = \binom{c+r-1}{c-1}\theta^c(1-\theta)^r, \quad r = 0, 1, \ldots, N-c. \tag{1.3a}$$

Similarly, for accepted batches, the sample size ranges from $(N-c+1)$ to $N$, with probabilities

$$\Pr\{n = N-c+1+r\} = \binom{N-c+r}{r}(1-\theta)^{N-c+1}\theta^{r}, \qquad (1.3b)$$

for $r = 0, 1, \ldots, (c-1)$.

Combining these two portions the total distribution of sample size is obtained, and we notice that for both small and high values of $\theta$ it will be positively skew, see Fig. 1.4.

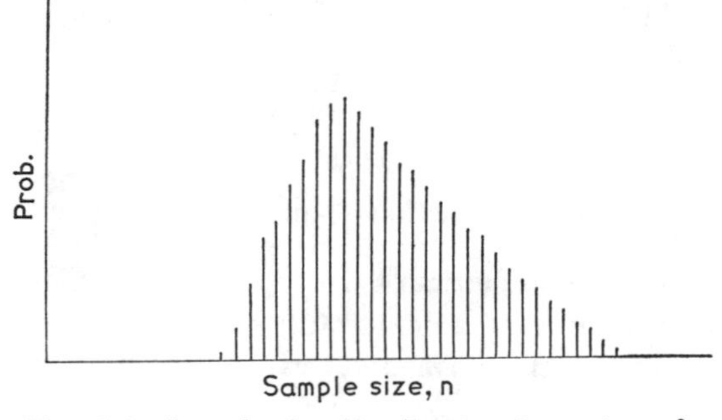


FIG. 1.4. Sample size distribution for a given $\theta$.

For some sequential plans, the sample size distribution may be difficult to evaluate and frequently it cannot be expressed in a concise form. In order to compare alternative boundaries, consideration is often limited to the average sample number curve (ASN-curve). This is a function of $\theta$ and for curtailed inspection it has a maximum somewhere near $\theta \simeq c/N$, see Fig. 1.5.

For small values of $\theta$ the ASN will be approximately $(N-c+1)/(1-\theta)$ and for high values of $\theta$, approximately $c/\theta$. The straight line on Fig. 1.5 represents the equivalent fixed sample size plan.

The main advantage of curtailed inspection can now be stated formally – that in comparison with an equivalent single sampling plan, curtailed inspection has the same OC-curve, but an ASN-curve nowhere higher and for some ranges of $\theta$ appreciably lower than the sample size of the single sampling plan.

A sampling inspection plan is defined by boundaries in the lattice diagram at which a batch is accepted or rejected. The curtailed inspection plan will, for example, reject when $c$ defectives have been found, regardless of whether the total number of items inspected is $c$ or $N$. It would seem to be more reasonable to reject at less than $c$ defectives when the sample size is small, and more when the sample

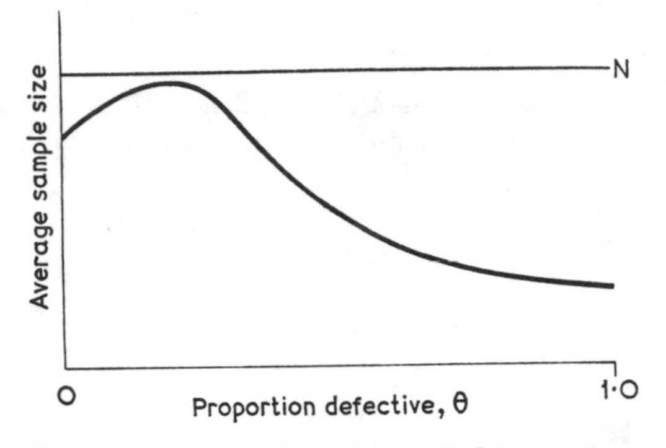


FIG. 1.5. ASN-function of curtailed inspection.

size is large. In fact boundaries of the type shown in Fig. 1.6a can be found which satisfy requirements (1.2) exactly. For low and high values of $\theta$ (but not at intermediate values) the ASN-curve is lower

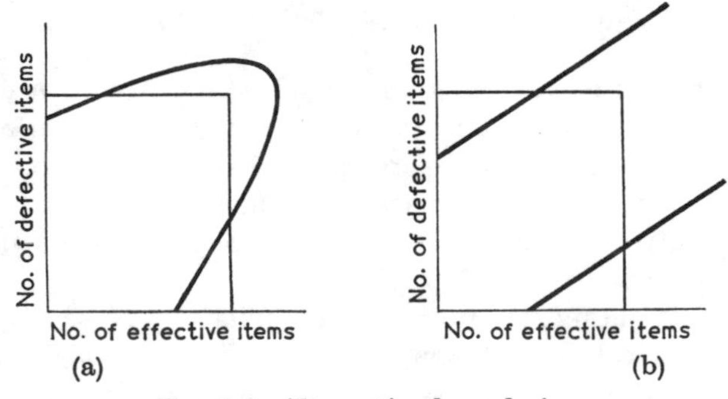


FIG. 1.6. Alternative boundaries.

than that given by the curtailed sampling plan satisfying requirements (1.2) but, except for the two points specified, the OC-curves are not exactly alike.

One possible theoretical approach is to ask what boundaries give some optimum property of the ASN-curve, while satisfying some restrictions on the OC-curve (such as 1.2). In the next chapter we show that Wald obtained the boundaries which satisfy (1.2), and also give the lowest possible ASN at both $\theta_0$ and $\theta_1$ (but not at intermediate values of $\theta$). Wald's boundaries are parallel lines, so that there is a positive (but rapidly decreasing) probability of obtaining a sample size larger than any given number, however large; these parallel boundaries often lead to a sample size distribution which is highly positively skewed.

Boundaries such as Fig. 1.6a, in which the sample size distribution has a finite upper limit, will be called closed boundaries, while those such as Fig. 1.6b, in which the sample size can be indefinitely large, will be called open.

### 1.4. Loss and Risk Functions

Consider the following example.

*Ex. 1.3.* Boxes of electronic components are available for being assembled into parts of a computer, and these components can be inspected and classified as either effective or defective. A sample from each box is tested before the box is used, and boxes are either passed or discarded. Every good component rejected costs a known amount; almost every bad item passed causes trouble at a later stage, and the cost of locating and eliminating the faulty item can be estimated. The testing of components is done on a machine one at a time, and the cost of this operation, in labour and material, is also known.

Suppose we wish to design a curtailed inspection plan for Ex. 1.3, then we could proceed by considering the OC-curve and ASN properties alone, but the final decision on a plan would be a matter of judgment, based on balancing various costs. In this example, the various costs involved are known reasonably accurately, and cost considerations can therefore be used explicitly in deciding on a sampling plan.

Denote the costs of accepting and rejecting a box having a percentage $\theta$ of bad items as $W_0(\theta)$, and $W_1(\theta)$ respectively; these are called decision loss functions. Suppose that the cost of inspecting items is independent of the number $n$ inspected from a box, and we shall take the cost of inspecting an item as the unit of costs. Denote the probabilities of accepting and rejecting a batch of quality $\theta$ under a sampling plan $S$ as $P_A(\theta|S)$, $P_R(\theta|S)$ respectively, where $P_A(\theta|S)+P_R(\theta|S)=1$. Then if $E(n|\theta,S)$ is the expected sample size, the risk function (expected loss) is

$$R(\theta|S) = E(n|\theta,S)+P_A(\theta|S)\, W_0(\theta)+P_R(\theta|S)\, W_1(\theta). \qquad (1.4)$$

If a risk function is known reasonably precisely, it should be considered, perhaps along with the OC- and ASN-curves, in choosing a plan. If a sampling plan existed for which the risk function were less than the risk for any other plan, for all values of $\theta$, then we should choose this plan. Such plans do not exist except in trivial cases such as when it is optimal not to inspect. We therefore need a method of choosing between alternative risk functions, and two principles have been suggested. One of these, proposed by Wald, is called the minimax principle. By this we choose the plan $S'$ which minimizes the maximum risk,

$$\underset{S}{\text{Min}}\ \underset{\theta}{\text{Max}}\ R(\theta|S).$$

This means that we are choosing a pessimistic plan – we are minimizing the worst possible risk. Frequently, better results are obtained with the minimax principle by choosing $S'$ for which

$$\underset{\theta}{\text{Max}}\{R(\theta|S) - \underset{S}{\text{Min}}\, R(\theta|S)\} = U(S)$$

is a minimum, which is called minimax regret, see § 7.7.

If we have a process curve or prior distribution of the frequency with which boxes of quality $\theta$ occur, then we can take expectations of the risk over this distribution of $\theta$ to obtain the Bayes risk $E\{R(\theta|S)\}$ which can be minimized with respect to the choice of $S$. Usually, there is insufficient knowledge of the process curve, so that this principle has to be regarded as approximate and also somewhat arbitrary.

These methods are considered in greater detail in Chapter 7. A plan obtained by minimizing expected loss is called a Bayes solution, and this method has the advantage that, if all costs involved are accurately represented, a unique optimum plan results, and the subjective judgment involved in choice of arbitrary restrictions such as (1.2) is removed.

**Problems 1**

1. Obtain a formula for the ASN of curtailed sampling, and show that it has a maximum near $c/n$.

2. Calculate the parameters of the curtailed inspection plans satisfying

$$\{\theta_0 = 0{\cdot}04 \quad \theta_1 = 0{\cdot}08, \quad \alpha = \beta = 0{\cdot}05\}$$

and

$$\{\theta_0 = 0{\cdot}045 \quad \theta_1 = 0{\cdot}075, \quad \alpha = \beta = 0{\cdot}04\}.$$

Plot the ASN- and OC-curves, roughly, for the two plans. What considerations determine which of these (and other similar) plans you would use in a particular application?

3. A single sample plan is being operated for acceptance inspection of large batches of items. Assume that you have back records of the number of defective items observed in samples of size 50, for 1,000 batches inspected in the previous three months. Consider under what conditions this information helps you to estimate the prior distribution, and how you would do it.

CHAPTER 2

# The Sequential Probability Ratio Test

## 2.1. Origin of the SPRT

Although sequential statistical methods were known for some time before, until World War II these were mainly very simple or *ad hoc* rules. The formal theory known today as 'Sequential Analysis' began in about 1943 with the work of A. Wald in America and G. A. Barnard in Britain in war-time industrial advisory groups. The most important discovery was Wald's 'Sequential Probability Ratio Test', and an elegant body of theory surrounding this was soon developed. This remains today a very important area of sequential analysis, and certainly an area in which the theoretical development is most complete.

The Sequential Probability Ratio Test, or SPRT, is well documented, see the book by Wald (1947), and a review and list of references in Johnson (1961). Below we give an outline of the theory, mostly without proofs, and for greater detail the reader is referred to these two references.

## 2.2. Derivation of the SPRT

The SPRT is designed to decide between two simple hypotheses. Suppose a random variable $x$ has a distribution $f(x, \theta)$, and we wish to test the null hypothesis that $\theta = \theta_0$ against the alternative hypothesis that $\theta = \theta_1$. The test constructed decides in favour of either $\theta_0$ or $\theta_1$ on the basis of observations (random variables) $x_1, x_2, \ldots$; we will suppose that if $\theta_0$ is true we wish to decide for $\theta_0$ with probability at least $(1-\alpha)$, while if $\theta_1$ is true, we wish to decide for $\theta_1$ with probability at least $(1-\beta)$.

For a fixed sample size (non-sequential) test, the optimum solution to this problem was provided by Neyman and Pearson (1928). They showed that for a given $n$, the test giving smallest $\beta$ (that is, the most powerful test) depends on the likelihood ratio $l_n$, where

$$l_n = \frac{\text{Probability of observed results given } \theta_1 \text{ true}}{\text{Probability of observed results given } \theta_0 \text{ true}}$$

$$= \prod_{i=1}^{n} \frac{f(x_i, \theta_1)}{f(x_i, \theta_0)},$$

and the test decides for or against $\theta_0$ according as $l_n$ is less than or greater than a constant. The value of this constant can be chosen to give the test the correct size $\alpha$, and in principle $n$ can be chosen to give the test power $(1-\beta)$.

Wald's SPRT is analogous to this, and has an analogous optimum property. It operates as follows.

Continue sampling as long as

$$B < l_n < A. \tag{2.1}$$

Stop sampling and decide for $\theta_0$ as soon as $l_n < B$, and stop sampling and decide for $\theta_1$ as soon as $l_n > A$. The constants $A$ and $B$ can be chosen to obtain approximately the probabilities of error $\alpha$ and $\beta$ prescribed, and the following theorem is relevant in this connection.

*Theorem 2.1* (*Wald*). For the sequential probability ratio test defined above, $A \simeq (1-\beta)/\alpha$ and $B \simeq \beta/(1-\alpha)$.

The proof of this theorem is along the following lines. There are four possible outcomes to any particular test:

| | *Outcome* | *Required Probability* |
|---|---|---|
| (i) | $\theta_0$ is true and the test decides in favour of $\theta_0$ | $(1-\alpha)$ |
| (ii) | $\theta_0$ is true and the test decides in favour of $\theta_1$ | $\alpha$ |
| (iii) | $\theta_1$ is true and the test decides in favour of $\theta_1$ | $(1-\beta)$ |
| (iv) | $\theta_1$ is true and the test decides in favour of $\theta_0$ | $\beta$ |

Denote the probability of an observed set of results $x_1, x_2, \ldots, x_n$, when $\theta_i$ is true as $p_i(x_1, x_2, \ldots, x_n)$, then for any set of results $x_1, x_2, \ldots, x_n$, leading to a decision in favour of $\theta_0$,

$$p_1(x_1, x_2, \ldots, x_n) < Bp_0(x_1, x_2, \ldots, x_n), \tag{2.2}$$

by the rule for termination.

Denote by $D_i$, for $i = 0, 1$, the set of all possible sets of results $(x_1, x_2, \ldots, x_n)$ for all $n$, which lead to a decision in favour of $\theta_i$, then by definition

$$\sum_{D_0} p_0(x_1, x_2, \ldots, x_n) = 1-\alpha, \tag{2.3}$$

$$\sum_{D_0} p_1(x_1, x_2, \ldots, x_n) = \beta, \text{ etc.} \tag{2.4}$$

(In words, expression (2.3) says that the sum of the probabilities of obtaining sets of results $\{x_1, x_2, \ldots, x_n\}$ given $\theta_0$ true, summed over all sets of results which terminate in a decision in favour of $\theta_0$, is the probability of deciding for $\theta_0$ when $\theta_0$ is true, or case (i) above.)

Thus from expression (2.2) we obtain

$$\sum_{D_0} p_1(x_1, x_2, \ldots, x_n) \leqslant B \sum_{D_0} p_0(x_1, x_2, \ldots, x_n),$$

or $B \geqslant \beta/(1-\alpha)$. The other inequality can be established similarly, $A \leqslant (1-\beta)/\alpha$. (Throughout this theorem we have assumed that a decision must eventually be made, see P2.1.)

Now since sampling is stopped as soon as either $l_n < B$ or $l_n > A$, the inequalities just given for $A$ and $B$ will usually be very nearly equalities. The errors involved in the approximations of Theorem 2.1 are determined by how much we overshoot the boundaries $B$ and $A$; Wald has shown that if $A$ and $B$ are chosen equal to the expressions of Theorem 2.1, no appreciable increase in either $\alpha$ or $\beta$ can result, and in fact, at most one of these quantities may be increased. Therefore, for most practical purposes, Theorem 2.1 provides a method of constructing a sequential test satisfying the requirements set out in the first paragraph of § 2.2.

A simple application of the SPRT is to sampling inspection of large batches of items, in which each item can be classified as effective or defective, and the batches are to be accepted or rejected. Suppose we require the sampling plan to accept the batch with probability $(1-\alpha)$ when the proportion defective is $\theta_0$, and reject the batch with probability $(1-\beta)$ when the proportion defective is $\theta_1$ (that is, we select two points on the OC-curve). Denote the number of effective and defective items found up to any stage of sampling as $s$, $r$ respectively, then the SPRT is of the form

$$B < \left(\frac{\theta_1}{\theta_0}\right)^r \left\{\frac{(1-\theta_1)}{(1-\theta_0)}\right\}^s < A.$$

Using Theorem 2.1 and taking logarithms we have a simple scoring procedure defined below (see Barnard, 1946),

$$\log\{\beta/(1-\alpha)\} < r\log(\theta_1/\theta_0) + s\log\{1-\theta_1)/(1-\theta_0)\} < \log\{(1-\beta)/\alpha\}, \quad (2.5)$$

which approximately satisfies the requirements.

This sampling plan has been advised in a number of sampling inspection tables, see Hill (1962) and references. However, before applying a plan of this type we would want further information in terms of the ASN-curve, further details of the OC-curve, and information about any optimum properties. We turn to these questions in the next section.

## 2.3. Some Important Properties of the SPRT

*(a) The OC-curve*

*Theorem 2.2 (Wald).* For the SPRT defined in Theorem 2.1 the OC-curve is approximately

$$P(\theta|\alpha,\beta,\theta_0,\theta_1) \simeq \frac{A^{h(\theta)}-1}{A^{h(\theta)}-B^{h(\theta)}} \quad (2.6)$$

where $P(\theta|\alpha,\beta,\theta_0,\theta_1)$ is the probability of deciding for $\theta_0$, and $h(\theta)$ is the solution of

$$\sum_x \left[\frac{f(x,\theta_1)}{f(x,\theta_0)}\right]^{h(\theta)} f(x,\theta) = 1 \quad \text{or} \quad \int_{-\infty}^{\infty} \left[\frac{f(x,\theta_1)}{f(x,\theta_0)}\right]^{h(\theta)} f(x,\theta)\,dx = 1 \quad (2.7)$$

depending on whether $f(x,\theta)$ is discrete or continuous, respectively. (We shall often simply write $P(\theta)$ for the OC-curve.)

As an illustration of the theorem, consider the binomial SPRT given by inequality (2.5). The OC-curve is

$$P(\theta) \simeq \left[\left(\frac{1-\beta}{\alpha}\right)^h - 1\right] \Big/ \left[\left(\frac{1-\beta}{\alpha}\right)^h - \left(\frac{\beta}{1-\alpha}\right)^h\right], \quad (2.8)$$

where $h(\theta)$ is the solution of

$$\theta\left(\frac{\theta_1}{\theta_0}\right)^h + (1-\theta)\left(\frac{1-\theta_1}{1-\theta_0}\right)^h = 1, \quad (2.9)$$

see Wald (1947, p. 51) where he suggests solving this equation for $\theta$ for given $h$ values, rather than vice versa.

Wald (1947) has shown how Theorem 2.1 can be used to prove Theorem 2.2; here we merely indicate the logical steps involved. Suppose we can find a function $h(\theta)$ such that

$$\int_{-\infty}^{\infty} \left[\frac{f(x, \theta_1)}{f(x, \theta_0)}\right]^{h(\theta)} f(x, \theta)\, dx = 1,$$

then the function

$$f^*(x, \theta) = \left[\frac{f(x, \theta_1)}{f(x, \theta_0)}\right]^{h(\theta)} f(x, \theta)$$

can be treated as a distribution function. The SPRT for testing $f^*(x, \theta)$ versus $f(x, \theta)$, for $\theta$ given, has the form

$$B_\theta^* < \frac{f^*(x, \theta)}{f(x, \theta)} < A_\theta^*. \tag{2.10}$$

If

$$A_\theta^* = A^{h(\theta)} \tag{2.11}$$

and

$$B_\theta^* = B^{h(\theta)}, \tag{2.12}$$

then it is easily seen that (2.1) and (2.10) are identical, and (2.10) is merely (2.1) to the power $h(\theta)$. That is, for any given sequence of observations $x_1, x_2, \ldots$, the SPRT defined by (2.1) stops sampling and decides for $\theta_0$ or $\theta_1$, at exactly the same point at which the SPRT defined by (2.10) stops and decides for $f(x, \theta)$ or $f^*(x, \theta)$ respectively.

Now suppose that $f(x, \theta)$ is the true distribution of $x$, then the probability that the SPRT defined by (2.1) decides for $\theta_0$ is $P(\theta)$, and by the above argument this is equal to the probability that the SPRT defined by (2.10) decides for $f(x, \theta)$, when this is the true distribution of $x$. But if we define for the SPRT (2.10),

$$\begin{aligned}
&\Pr\{f(x, \theta) \text{ true and SPRT (2.10) decides for } f(x, \theta)\} &&= 1-\alpha^* \\
&\Pr\{f(x, \theta) \text{ true and SPRT (2.10) decides for } f^*(x, \theta)\} &&= \alpha^* \\
&\Pr\{f^*(x, \theta) \text{ true and SPRT (2.10) decides for } f^*(x, \theta)\} &&= 1-\beta^* \\
&\Pr\{f^*(x, \theta) \text{ true and SPRT (2.10) decides for } f(x, \theta)\} &&= \beta^*,
\end{aligned}$$

then the above argument shows that

$$P(\theta) = 1-\alpha^*. \tag{2.13}$$

Now by applying Theorem 2.1 to the SPRT (2.10)

$$A_\theta^* \simeq (1-\beta^*)/\alpha^* \qquad (2.14)$$

$$B_\theta^* \simeq \beta^*/(1-\alpha^*). \qquad (2.15)$$

By using equations (2.11) to (2.15), Theorem 2.2 is obtained.

In Wald's derivation of the OC-curve (and of the ASN-function), an identity known as the fundamental identity plays an important part. Wald (1947) proved that

$$E\{e^{Z_n t}[\phi(t)]^{-n}\} = 1,$$

where

$$\phi(t) = E(e^{zt}),$$

and

$$z = \log\{f(x,\theta_1)/f(x,\theta_0)\},$$

and where $Z_n$ is the score when a decision is taken,

$$Z_n = z_1 + z_2 + \ldots . z_n.$$

This identity holds under very general conditions, and not only for the SPRT. For a generalization of Wald's identity and references to important related papers see Miller (1961). The proof of the OC-curve formula given above is also due to Wald, and it is given here because it is more suitable for the discussion of extensions to Wald's theory by Bartholomew (1956); see Chapter 5.

*(b) The ASN function*

One striking feature of the SPRT is that, in spite of the mathematical complexity with which most investigations in sequential analysis are fraught, some of the main results regarding properties of the SPRT are surprisingly simple provided the observations $x_i$ are independent. Wald derives an approximation to the ASN-function as follows. Take logarithms of (2.1) and we have

$$\log B < \sum z_i < \log A \qquad (2.16)$$

where

$$z_i = \log\{f(x_i,\theta_1)/f(x_i,\theta_0)\}.$$

This is a scoring procedure with independently distributed scores $z_i$, and, neglecting the overshoot of the boundaries, the total score when a decision is made will be $\log B$ or $\log A$, with probabilities $P(\theta)$, and $\{1-P(\theta)\}$ respectively, by Theorem 2.2.

Let $n$ denote the sample size required for a decision, and write the probability that $n$ takes any value $r$ as $p(r|\theta)$, where $r = 1, 2, \ldots$ Associate with every random variable $z_i$, where $i = 1, 2, \ldots$, a variable $\delta_i$, which is unity if a decision has not been reached before sample size $i$, and zero otherwise. Write the score when a decision is taken as $Z_n$, so that for the SPRT, $Z_n$ takes on two values,

$$\Pr\{Z_n = \log B\} = P(\theta)$$

and

$$\Pr\{Z_n = \log A\} = 1 - P(\theta).$$

Now

$$Z_n = z_1 + \ldots + z_n = \sum_1^\infty \delta_r z_r$$

and on taking expectations

$$E(Z_n) = \sum_1^\infty E(\delta_r z_r) = \sum_1^\infty E(z_r)\, E(\delta_r)$$

since $\delta_r$ depends only on $z_1, \ldots, z_{r-1}$, and not on $z_r$. If the $z_i$ are identically and independently distributed,

$$E(Z_n) = E(z) \sum E(\delta_r) = E(z) \sum_1^\infty rp(r|\theta) = E(z)\, E(n|\theta). \quad (2.17)$$

This equation is valid for a broad class of sequential tests. By inserting our knowledge of $E(Z_n)$ for the SPRT we have Theorem 2.3.

*Theorem 2.3.* An approximation to the ASN-curve of the SPRT defined by (2.1), with probabilities of error $\alpha$ and $\beta$ is

$$E(n|\theta) \simeq \{P(\theta) \log B + (1 - P(\theta)) \log A\} / E(z|\theta). \quad (2.18)$$

We notice that the approximation breaks down for $E(z|\theta) = 0$, and Wald obtains a special formula for this case. Two approximations are involved in the use of Theorem 2.3, firstly the error resulting from ignoring the overshooting of the boundaries, and secondly the approximation involved in $P(\theta)$ from Theorem 2.2. Wald was able to obtain limits to the ASN-function, which provides a partial check of the approximations. The next section refers to some empirical sampling trials which provide a better check of Theorem 2.3 for a particular application.

Now if the distribution of $z_i$ in (2.16) is $\phi(z, \theta)$, it is easily seen that, for a starting point $Z$, and given $\theta$, the expected sample size satisfies the equation

$$E(n|Z, \theta) = 1 + \int_{\log A}^{\log B} E(n|y, \theta)\, \phi(y - Z, \theta)\, dy. \qquad (2.19)$$

A similar integral equation (without the constant unity on the right-hand side), holds for the OC-curve, and Page (1954) and Kemp (1958) have developed approximate methods of solving these equations, which usually yield better results than Wald's approximations, see the source papers for details. A number of other approximations to the OC and ASN curves of the SPRT have been developed, which are better than the approximations given by Wald in certain circumstances; see Manly (1969) and the references therein for some of this work.

Raghavachari (1965) has evaluated the exact OC and ASN functions in the case where observations are exponentially distributed, and he also gives a comparison between the exact functions and the Wald approximations.

*(c) The sample size distribution*

The SPRT being an open sequential test, we might expect the distribution of sample size to be skew, and two sources of evidence confirm this.

Firstly some asymptotic properties of the sample size distribution were obtained by Wald, and some exact results in special cases; for the most part the theoretical results are not very illuminating; see Johnson 1961, p. 383, for a short discussion and list of references on this point.

Secondly, Baker (1950) carried out some empirical sampling trials, for a test of a normal mean, $H_0: N(0,1)$ versus $H_1: N(1,1)$, and four sets of errors $\alpha$ and $\beta$. For $\alpha = \beta = 0{\cdot}05$ the ASN predicted by Theorem 2.3 at the null hypothesis was 5·6 as against an observed 7·0; out of 2,003 SPRT's, 189 had sample sizes of 14 or more and 56 sample sizes of 19 or more. For the four sets of SPRT's the lowest frequency of sample sizes more than three times the average was about 2%, and the lowest frequency of sample sizes more than twice the average was about 10%.

Very similar results were obtained by Corneliussen and Ladd (1970) for tests of the binomial probability $\theta$. These authors developed a recursive method for exact calculation of the sample size distribution and ASN. Probability contours of the sample size distribution are given, and these demonstrate quite clearly the extreme skewness of this distribution. Wald's formula for the ASN is shown to under-represent the true ASN at the maximum by about 20% and the authors conclude, '. . . in view of the very large spread in sample number, we do not believe the ASN function to be a particularly useful measure of the effort required to conduct a sequential test.'

Thus the saving in observations resulting from use of an SPRT is very much an average property, and in particular cases, an SPRT may chance to require many more observations than a non-sequential plan having the same probabilities of error.

Baker's results also provide a check on the accuracy of Theorem 2.1 and 2.3 approximations. The actual probabilities of rejecting the null hypothesis were about 30% less than the values of $\alpha$ used in Theorem 2.1. This result might be expected since any overshooting of the boundaries will presumably lead on average to operating with greater safety than set.

The work of Baker (1950), Page (1954), Kemp (1958) and Corneliussen and Ladd (1970) regarding the ASN all indicate that the approximation in Theorem 2.3 can sometimes substantially under-estimate the true ASN, especially if the starting point of the SPRT is close to one of the boundaries.

Finally, Cox and Roseberry (1966a) present some empirical evidence to show that the variance of the sample number is approximately proportional to the square of the ASN. This bears out the remarks made above about the skewness of the sample size distribution.

*(d) An optimum property*

Wald and Wolfowitz (1948), see also Wolfowitz (1966) proved a property of the SPRT, which is very interesting from a theoretical point of view.

*Theorem 2.4.* For all sequential tests of $f(x, \theta_0)$ against $f(x, \theta_1)$, having probabilities of error $\alpha$ and $\beta$, the SPRT has the least possible values of $E(n|\theta_0)$ and $E(n|\theta_1)$.

Therefore if the only possible values of $\theta$ are $\theta_0$ or $\theta_1$, and if our requirements for a sequential test can be reduced to specifying probabilities of error $\alpha$ and $\beta$ at those values of $\theta$, then the SPRT produces the lowest possible ASN at both $\theta_0$ and $\theta_1$. If Theorem 2.1 approximations are used to construct a test, the property only holds roughly. Further, although the ASN will be approximately optimum at (or near) $\theta_0$ or $\theta_1$, we shall show that at intermediate values of $\theta$ $E(n|\theta)$ may even be greater than the sample size of an equivalent non-sequential plan with the same probabilities of error.

In the preceding section we have discussed results which indicate that the sample size distribution is heavily positively skewed, and it seems clear that Theorem 2.4 is obtained at the cost of a rather undesirable sample size distribution. Suppose an SPRT is being used in a sampling inspection plan, then although the SPRT will probably lead to a sizeable saving in inspection when averaged over six months' work, this may not compensate for regular bottle-necks caused by large sample sizes. It may be doubted whether situations often arise in which we want to minimize the ASN without regard to other features of the sample size distribution.

One possibility discussed by Wald was to truncate the SPRT at some sample size $n_0$, and this is no doubt a common practice. Wald produced a method of choosing $n_0$ large enough to have a negligible effect.

One field in which unrestricted minimization of the ASN may be applicable is computer simulation trials. Here we may have a large number of tests, and occasional large samples will not matter.

For truncated SPRTS, Aroian (1968), and Aroian and Robison (1969), have shown how exact OC and ASN curves can be calculated. For continuous distributions the method used is simply to work from one sample size to the next by numerical integration so as to determine the probabilities of either type of decision, and the p.d.f. in the continuation region.

It is worth pointing out here that it has been assumed throughout the above discussion that the non-truncated SPRT terminates with probability one. For proofs of this see Wald (1947), Ghosh (1970), Ifram (1965), and Wijsman (1967).

It will assist us to evaluate the SPRT if we consider some applications in detail.

## 2.4. Some Applications of the SPRT

*Ex. 2.1.* Suppose we wish to carry out a binomial SPRT (2.5), for $\theta_0 = 0{\cdot}25$ versus $\theta_1 = 0{\cdot}75$, having probabilities of error $\alpha = \beta = 0{\cdot}001$. By using equations (2.8), (2.9), (2.18) and a special formula given by Wald, (1947, p. 99) for the case $E(z|\theta) = 0$, we obtain Table 2.1.

TABLE 2.1. ASN for the binomial SPRT of Ex. 2.1.

| $\theta$ | 0·25 | 0·37 | 0·43 | 0·50 | 0·57 | 0·63 | 0·75 |
|---|---|---|---|---|---|---|---|
| ASN | 12·6 | 22·0 | 32·2 | 39·5 | 32·2 | 22·0 | 12·6 |

It is easily shown that for a single sampling plan using thirty-three observations, the probabilities of error are 0·00095 (accept $\theta_1$ for seventeen or more positive responses). From Table 2.1 we see that there is a good range of values of $\theta$ near $\frac{1}{2}$ for which the ASN of the SPRT is greater than the equivalent fixed sample size, and the discussion of § 2.3 indicates that the exact ASN may be slightly greater than the values given in Table 2.1. However, there is a considerable saving in observations close to $\theta_0$ and $\theta_1$, and Theorem 2.4 proves that the ASN at these values of $\theta$ are (approximately) as low as they can be for the given probabilities of error.

In many SPRT it happens that the ASN-curve is lower than the equivalent fixed sample size throughout, but the maximum of the ASN-curve frequently represents a rather small saving in observations, especially if the probabilities of error are small.

We therefore conclude this discussion as follows. An SPRT is obtained by performing a test between two hypothetical values of an unknown parameter (or set of parameters); if the values of the unknown parameter occurring in practice are most commonly close to the hypothesized values, the SPRT is probably very efficient in terms of average sample size. However, if the hypothesized values are taken merely to obtain an SPRT, and do not represent the most frequently occurring values, then the SPRT may not lead to much saving in observations, and occasionally leads to a net loss.

*Ex. 2.2* (*Oakland*, 1950). 'The whitefish, *Coregonus clupeaformis*, sometimes contains cysts of the tapeworm *Triaenophorus crassus*. The presence of these cysts in the edible parts of fish is obnoxious but not harmful to man. When whitefish are exported from Canada it is desirable to know the infestation rate of the lot. This infestation rate is computed in the following manner: a certain number of fish are selected from the lot; they are examined for the number of cysts and the infestation rate is obtained from the formula:

$$\text{I.R.} = \frac{\text{Total number of cysts in sample}}{\text{Total weight of fish in sample in lb}} \times 100.$$

If the infestation rate is higher than a certain number, the lot is not suitable for export. If the infestation rate is lower than a certain number, the fish are exported. The test described above is a destructive test.'

Oakland explained that the infestation rate of loads of fish varied a good deal, and in order to ensure sufficient high quality, yet with minimum loss to the shipper, a sequential sampling plan was a natural choice. Oakland's description of the derivation of the boundaries is along the following lines.

He chose two hypotheses, $H_0$ that the infestation rate is 10 cysts per 100 lb of fish, and $H_1$ that it is 30 cysts per 100 lb of fish. It is easier to operate a sequential sampling plan in terms of cysts per fish rather than infestation rates, and these hypotheses are converted to cysts per fish. For fish in the round medium range (1·5 to 3 lb), working with an average of 2 lb per fish yields means of 0·2 and 0·6 cysts per fish for $H_0$ and $H_1$ respectively.

Oakland examined some data on the distribution of cysts per fish and found this to be closely negative binomial,

$$f(x, \theta) = \theta^{-\kappa} \frac{\Gamma(x+\kappa)}{\Gamma(\kappa)\, x!} \left(\frac{\theta}{1+\theta}\right)^{\kappa+x} \tag{2.20}$$

where $x = 0, 1, 2, \ldots$ The observed mean and variance were 1·1 and 2·1 respectively, for round medium fish. The hypotheses $H_0$ and $H_1$ are therefore represented by sets of parameters $(\kappa_i, \theta_i)$, $i = 0, 1$. The alternative hypothesis was chosen to have a mean of 0·6 and a variance of 2 (the latter being close to the observed variance, and the former, while also being close to the observed mean, is equivalent to the desired infestation rate). This yields for $H_1$

$$\begin{aligned} \kappa_1 \theta_1 &= 0{\cdot}6 & \kappa_1 &= 0{\cdot}2571 \\ \kappa_1 \theta_1(1+\theta_1) &= 2 & \theta_1 &= 2{\cdot}3333. \end{aligned}$$

For reasons largely of mathematical convenience, Oakland used the same $\kappa$ for both hypotheses. In order to obtain the desired mean of 0·2 for $H_0$, we therefore have

$$\kappa_0 = 0{\cdot}2571 \qquad \theta_0 = 0{\cdot}7779.$$

Oakland proceeded to obtain an SPRT for testing these two hypotheses, with $\alpha=\beta=0{\cdot}05$, and he also examined the OC-curve and the ASN-curve. The reader is encouraged to work through these details himself and check his results with those given in Oakland's paper.

The choice of the two hypotheses appears to be rather arbitrary, and in no sense represents most frequently occurring sets of parameters. The SPRT here is a convenient method of obtaining a sequential test, for which simple approximations to the OC- and ASN-curves are available; the properties of the test obtained can presumably be worked out and checked to see if they are reasonable.

Since the quality presented for inspection is so variable, a sequential test is a good choice for this application and the SPRT no doubt pays off considerably, although no optimum property could be claimed for it. Any fixed sample size plan would have to have a large sample size in order to achieve the desired error rates, yet with very good or very bad carloads of fish a very small sample would be sufficient.

Oakland reported one practical difficulty which often arises. Sometimes the fish are selected and inspected in places separated by some short distance, and it becomes impossible to keep going back for more samples. One way of circumventing this difficulty is to perform the test in groups of a given size. This must increase the ASN, but the plan would be much easier to operate.

There is a further difficulty here, not dealt with by Oakland. The final SPRT depends on the observations being distributed according to the negative binomial distribution having $\kappa=0{\cdot}2571$. Variations in $\kappa$ will alter the properties of the SPRT, and the effect of such variations does not appear to have been investigated.

*Ex. 2.3.* Morgan *et al.* (1951) have described an application of the SPRT to the grading of milk in the state of Connecticut.

Films of milk were prepared and viewed under a microscope. A number of microscopic fields were observed on each film and the number of bacterial clumps counted. The state laws required that when using a microscopic factor of 600,000, thirty fields be examined and there be less than 200,000 bacterial clumps per ml for grade A milk.

Morgan *et al.* proceeded to prepare a number of films from various samples of milk, and then counted a large number of fields in each film. By examining the data from these samples they detected a lack of randomness of bacteria on a slide, there being a tendency to fewer clumps along the borders. A sampling method is suggested to reduce possible bias due to this effect. The distribution of the number of clumps per field was found to agree very well with the Poisson distribution except for milk films with very low or very high counts. A comparison of different films from the same milk sample also gave significant results only for very low or very high count milks, so that reliance upon a single film would seldom lead to a major error.

Morgan *et al.* therefore assumed a Poisson distribution of clumps per field, and calculated an SPRT giving an OC-curve close to that of the official method (the curves were made to agree at the 5% probability of error points). The ASN-curve showed that there was always a substantial saving of inspection over the official method.

An empirical study of the suggested SPRT was carried out as follows. Series of 100 separate fields were counted on each of a number of films (actually the data obtained for checking the distribution assumptions was used). Using these observations, SPRT were simulated, and from one to eleven SPRT's were obtained from each set of 100 field counts. The 'true' count was taken to be the average of the 100 counts on each film, and the percentage of SPRT giving grade A, and the ASN were plotted against these estimated true values. (Actually the results were grouped according to 'true' value before plotting.) The OC-curve obtained from these results appeared to give somewhat lower probabilities of error than those set, and the ASN-curve was slightly higher than the theoretical approximation obtained from Theorem 2.3.

Ex. 2.3 is the most satisfactory published application of the SPRT I have been able to find. Only in two other cases have I found any evidence of checking of the assumptions and empirical checking of the final plan; one of these cases is a study by Anastasi (1953), which forms a basis for the application of the SPRT to the selection of items for psychological tests, see Burgess (1955). The other application including empirical checking of the final plan, etc., is by Waters (1955), briefly described in Chapter 3.

The references list a number of practical applications. The expository articles all give details of how the SPRT could be used in specific fields, but it is not clear whether the SPRT had actually been employed.

Rivett (1951) discussed machine process setting, in which effectively a normal mean with known variance had to be kept within tolerance limits $(m_L, m_U)$. Rivett's suggestion was to carry out an SPRT to test the hypothesis that the mean is $m_L$ against the alternative hypothesis that it is $m_U$; if no decision has been reached by a certain sample size, inspection is stopped and the mean is assumed to be between $m_L$ and $m_U$. This use of the SPRT seems almost to maximize the sample size ordinarily used, and a more appropriate formulation can be given in terms of Chapter 3 methods.

Cowden's (1946) application was to the grading of students based on the scripts of a statistics examination. The application is an interesting one, but it seems doubtful whether the underlying assumptions of the SPRT are fulfilled.

Jones (1947) used the SPRT to check the percentage of errors made by individuals punching U.S.A. Social Security numbers on Hollerith cards. A high percentage of errors was traced to bad ventilation, defective equipment, a need for better instruction, etc. When errors arise, they most probably are not random, and do not follow any distribution law. Nevertheless, the technique of using the SPRT was probably a good one.

Enrick (1946) described the U.S. Army Quartermaster Corps' use of the SPRT, and this is probably typical of a number of applications to sampling inspection. Sampling was carried out in groups, and the plan truncated after a few groups. The parameters of the SPRT were chosen in a rather arbitrary manner, the formulation of § 2.2 being viewed merely as a technique for obtaining a simple sequential plan having (roughly) an OC-curve which seems desirable.

Wald realized that a formulation of sequential tests as tests between two simple hypotheses, $\theta = \theta_0$, and $\theta = \theta_1$, is totally unrealistic, and he discussed a formulation in which $(\theta_0, \theta_1)$ is an indifference region in a choice between two decisions (see § 1.3, and Wald, 1947, p. 70). However, even in this revised form, the SPRT formulation appears unrealistic, and it does not seem to fit any of the applications studied. Usually $\theta_0$ and $\theta_1$ are taken to be extremes, rather than bordering values of an indifference region.

Johnson (1960) suggested that the SPRT could be designed as follows. If an SPRT is constructed using two values of $\theta$, say $(\theta_0', \theta_1')$, and two corresponding probabilities of error $(\alpha', \beta')$, the OC-curve is given by Theorem 2.2 and is $P(\theta|\alpha', \beta', \theta_0', \theta_1')$. Suppose we require the OC-curve to go (nearly) through two points specified by $(\theta_0, \theta_1)$ and $(\alpha, \beta)$, then we must satisfy

$$P(\theta_0|\alpha', \beta', \theta_0', \theta_1') = 1-\alpha$$

and

$$P(\theta_1|\alpha', \beta', \theta_0', \theta_1') = \beta.$$

Thus, for example, if $(\theta_0, \theta_1)$, $(\theta_0', \theta_1')$, and $(\alpha, \beta)$ are chosen, these equations fix $(\alpha', \beta')$. There is no need to choose $\alpha = \alpha'$, $\theta_0 = \theta_0'$, etc.; $(\theta_0, \theta_1)$ with corresponding probabilities of error $(\alpha, \beta)$ could be chosen to correspond to be rather extreme positions on the OC-curve, while $(\theta_0', \theta_1')$ represent frequently occurring values of $\theta$. This should lead to a more satisfactory ASN-function, but Johnson notes from some calculations that 'the approach is not spectacularly successful'.

**Problems 2**

1. Prove that for independent observations the SPRT defined by (2.1) terminates with probability one. (See Wald, 1947, p. 157.)

Readers will notice that Theorem 2.1 is valid for non-independent observations provided the process terminates with probability one.

2. Obtain an SPRT for hypergeometric sampling, comparable with the binomial SPRT of equation (2.5). Use Stirling's approximation to simplify your procedure. Compare the boundaries with those for binomial sampling.

3. Suppose independent random variables $x_i$ can be observed, which are normally distributed with $\sigma^2 = 1$, and mean $\theta$ unknown. You are required to test $H_0: \theta = -d/2$ against $H_1: \theta = +d/2$, with probabilities of error $\alpha = \beta$, where $\theta$ is the unknown mean.

Obtain (a) the fixed sample size test, and (b) the SPRT, and its OC- and ASN-curves. Show that at $\theta = 0$, there is a range of $\alpha$ for which the ASN of the SPRT is greater than the fixed sample size, independently of $d$. Find the value of $\alpha$ for which the ASN of the SPRT is equal to the fixed sample size.

Comment on the practical importance of the result.

4. Empirical sampling trials have shown that the ASN given by Theorem 2.3 is up to 30% an underestimate of the true ASN. Repeat the calculations of P2.3 assuming that the ASN should be inflated by $p$%, for various values of $p$ up to 30%. Also draw the ASN-function for P2.3, for $\alpha = \beta = 0{\cdot}05$, assuming that the true ASN is 25% more than the theoretical estimates.

Comment on the implications of these results to considerations of when you would advise using the SPRT.

5. In Ex. 2.2, an SPRT has been obtained for the mean of a negative binomial distribution, assuming $k$ known. Outline an investigation into the effect on the properties of the resulting test if the constant $k$ is in error by varying amounts.

6. For the SPRT of a normal mean $\theta$ when the variance $\sigma^2$ is known, consider how the likelihood of $\theta$ varies along each stopping boundary. If it is agreed that inferences from data are to be made on the likelihood alone (see Barnard *et al.*, 1962), consider the implications of this variation of the likelihood.

7. Schneiderman and Armitage (1962a, p. 53, Table 6) give the results of some empirical sampling trials of a normal mean, $\sigma^2$ known. They use $H_0: \theta = 0$, $H_1: \theta = 1$, $\alpha = \beta = 0{\cdot}05$, and the observed ASN- and OC-curves are tabulated. Compare these with the theoretical ASN- and OC-curves, and with the ASN- and OC-curves for a suitably chosen fixed sample size plan.

8. Suppose that independent pairs of observations, $(x_i, y_i)$ can be observed, where each pair is drawn from a bivariate normal population with zero means, marginal variances $\sigma_x^2$, $\sigma_y^2$, and a correlation coefficient $\rho$: Obtain sequential tests for $H_0$: $\rho = \rho_0$ against $H_1$: $\rho = \rho_1$ in the following cases:

(a) When both $\sigma_x^2$ and $\sigma_y^2$ are known.

(b) When only the ratio $\sigma_x^2/\sigma_y^2$ is known by considering the sequence

$$V_i = \sigma_x y_i / \sigma_y x_i,$$

for $i = 1, 2, \ldots$

Obtain and compare the properties of the tests (see Kowalski, 1971). A test for the case where both $\sigma_x^2$ and $\sigma_y^2$ are unknown is given by Choi (1971).

9. Obtain an SPRT for testing $H_0$ that observations are independently distributed according to a standard normal distribution, against $H_1$, that observations are independently distributed with a Cauchy distribution, with a p.d.f.

$$\frac{\lambda}{\pi(1+\lambda^2 x^2)}, \qquad -\infty < x < \infty, \lambda > 0,$$

where $\lambda$ may be assumed known.

10. Consider the situation leading to the sequential test (2.5), and obtain a sequential test applicable to the case where items can only be sampled in groups of $k$ at a time. Derive the properties of your test.

11. A system is either 'active' or 'under repair' and active and repair times are independently distributed with p.d.f.'s

$$\exp(-x/\theta)/\theta \qquad \text{and} \qquad \exp(-y/\phi)/\phi$$

respectively. System availability is defined as

$$A = \theta/(\theta+\phi)$$

and it is required to test $H_0$: $A = A_0$ against $H_1$: $A = A_1$, with probabilities of error $(\alpha, \beta)$. Show that

$$\Pr(Y \leq X) = \theta/(\theta+\phi)$$

and hence construct a sequential test by using the sequence

$$z_r = \begin{cases} 1 \text{ if } y_r \leq x_r \\ 0 \text{ if } y_r > x_r \end{cases}$$

Comment on the efficiency of the test procedure (see Schafer and Takenaga, 1972).

CHAPTER 3

# Sequential Tests Between Three Hypotheses

## 3.1. Two Examples

So far we have described a theory of sequential tests suitable for discriminating between two simple hypotheses, and applications have been described in which discrimination between two courses of action is required, as in the whitefish sampling example, Ex. 2.2.

There are many practical situations where a choice among three or more courses of action is required, and the theory of the previous chapter is not appropriate. For example, the application to setting a machine within tolerance limits, suggested by Rivett (1951), required a choice among the possibilities that the setting was too high, too low, or acceptable (see brief discussion of this example in § 2.4). Also consider the following examples.

*Ex. 3.1.* Anderson (1954) described a simple method of comparing the strength of two yarns. Instead of estimating the average strength of each yarn, and comparing these estimates, Anderson proceeded as follows to estimate $\Pr(X > Y)$, where $X$ and $Y$ are random variables representing the strength of random lengths from the two yarns. Lengths of approximately 1 ft from each yarn were either knotted together or had their ends clamped together in a light grip. The yarns were then pulled until one yarn broke and it was noted in which yarn the break occurred. This was repeated until it was evident either that both yarns were of approximately equal strength, or that one or other was stronger. Anderson gave a sequential plan, based on Armitage (1950), for choosing between these three terminal decisions.

*Ex. 3.2.* Waters (1955) briefly described the general application of sequential schemes to various survey problems in forestry aimed at control of losses by destructive forest insects. Much of the burden of the surveys falls on a (U.S.A.) Federal forest-insect-survey organization, which is limited in finance and manpower; therefore any method of reducing the average amount of inspection involved is worth while.

He went on to describe in detail a sequential procedure for inspection of spruce budworm larvae. The insect larvae infestation was to be classified as light, medium or heavy, by counting the numbers of larvae on twigs 15 in. long cut by pole-pruner from mid-crown height of trees. The standard (fixed sample-size) procedure required five twigs from each of five trees at a number of collection points. Waters gave new sequential plans for classification of the insect larvae infestation, and it was briefly compared with the standard procedure in a field trial.

Ex. 3.1 and Ex. 3.2 both describe situations where the theory given in the previous chapter cannot be applied directly, and therefore some modification of this theory must be made or else new methods developed. Both examples could be formulated in terms of a sequential test among three hypotheses.

In Ex. 3.2 three simple hypotheses could be stated, representing light, medium and heavy insect infestations respectively; these three hypotheses are ordered in terms of the expected results from samples. Similar reasoning applies to Ex. 3.1 and this suggests that a satisfactory sequential test may be obtained by performing two SPRT's simultaneously, one between each pair of neighbouring hypotheses in the ordering. This idea was proposed by Armitage (1947, 1950), and Sobel and Wald (1949), and they show how to obtain sequential tests which approximately satisfy certain restrictions on the probabilities of misclassification.

### 3.2. Formulation of the Test

Consider the following statistical model for Ex. 3.1. Suppose that the two yarns being tested are $A$ and $B$, and suppose that the probability that yarn $A$ will break first is constant from trial to trial, and independent of other trials; denote this probability by $\theta$.

It is required to discriminate between the three possible situations, $A$ is stronger than $B$, $A$ and $B$ are equally strong, and $A$ is weaker than $B$. These three situations are represented by three simple hypotheses about $\theta$,

$$H_{-1}:\theta = \theta_1 < \tfrac{1}{2}, \qquad H_0:\theta = \tfrac{1}{2}, \qquad H_1:\theta = 1-\theta_1. \tag{3.1}$$

An SPRT $S_{-1}$ can be constructed to discriminate between $H_{-1}$ and $H_0$ with probabilities of error

$$\left.\begin{aligned}\Pr\{\text{accept}\, H_{-1}|H_0\,\text{true}\} &= \alpha/2\\ \Pr\{\text{accept}\, H_0|H_{-1}\,\text{true}\} &= \beta,\end{aligned}\right\} \tag{3.2}$$

and similarly a test $S_1$ to discriminate between $H_0$ and $H_1$, with probabilities of misclassification obtained by replacing $H_{-1}$ in (3.2) by $H_1$. These SPRT's can be represented in the lattice diagram as shown in Fig. 3.1, where $y$ is the number of breaks in yarn $A$, and $x$ the number of breaks in yarn $B$.

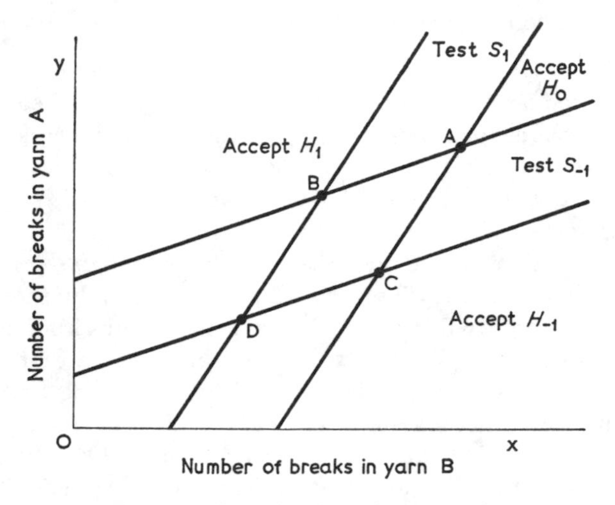


FIG. 3.1. Combination of two SPRT's.

For a given number $n$ of pairs of lengths of yarn tested, the results must be one of the points on the line $y+x=n$. Provided $n$ is large enough, the preferences given by the two SPRT's along this line represent the natural ordering of the hypotheses; that is, for increasing values of $y$, given $y+x=n$, we reach a region in which $S_{-1}$ prefers $H_0$ to $H_{-1}$ while $S_1$ still prefers $H_0$, then for higher values of $y$, $S_1$ prefers $H_1$. However, if the sample size $n$ is small enough for the line $y+x=n$ to intersect with the decision boundaries before $BC$, no such consistent pattern results. For example at $D$, $S_{-1}$ prefers $H_{-1}$ to $H_0$ and simultaneously $S_1$ prefers $H_1$ to $H_0$. The points $B$ and $C$ are possible observed points if the risks (3.2) are such that $\alpha > 2\beta$. Armitage (1947) and Sobel and Wald (1949) effectively insert conditions so that $BC$ is not in the positive quadrant; Armitage (1950) has a better way round the difficulty, described later. For the present, suppose that neither $B$ nor $C$ is in the positive quadrant, and we obtain Fig. 3.2. The shaded areas are those in which both $S_{-1}$ and $S_1$ inequalities are violated, and there are only three possibilities, corresponding to the three different terminal decisions, as follows:

| *Test* $S_{-1}$ *Accepts* | *Test* $S_1$ *Accepts* | *Terminal Decision* |
|---|---|---|
| $H_{-1}$ | $H_0$ | $H_{-1}$ |
| $H_0$ | $H_0$ | $H_0$ |
| $H_0$ | $H_1$ | $H_1$ |

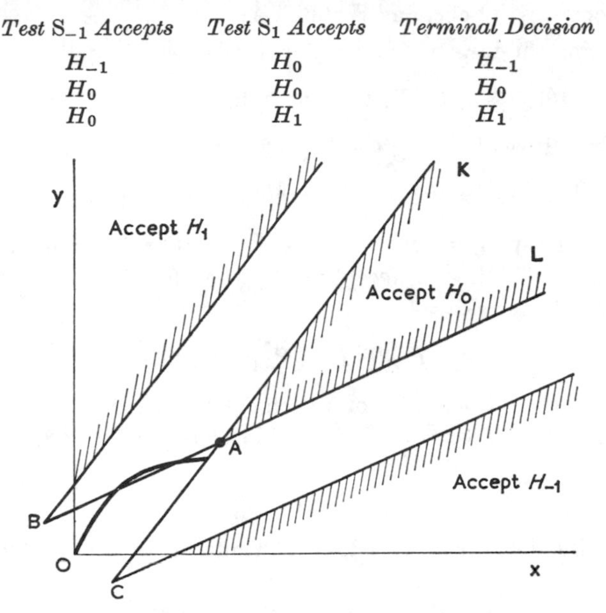


FIG. 3.2. Scheme for test of three hypotheses.

One procedure suggested is to operate $S_{-1}$ and $S_1$ simultaneously, and stop calculating each as soon as its own inequalities are violated, regardless of what has happened to the other test. Thus for the path shown in Fig. 3.2, $S_{-1}$ is not calculated after $AB$ has been crossed (even though the sample path may go back over the line later) and $S_1$ is stopped at $AC$ as shown. This path has led to a termination of the test procedure, before a shaded area has been reached. For such paths $H_0$ is accepted, since both tests have been terminated with a preference for $H_0$. Under this rule, therefore, the termination of the combined test depends not only on the total sample results, but also on the particular sample path chosen: Sobel and Wald point out that a test having this property cannot be optimum in any sense, since the test is not simply a function of the sufficient statistics, although the probability of termination before the shaded region $KAL$ is reached is sometimes small. The operation of this rule of terminating each SPRT independently of the other, leads to certain purely mathematical advantages which are fully exploited by Sobel and Wald.

### 3.3. Properties of Sobel and Wald's Test

Considering the tests $S_{-1}$ and $S_1$ separately, we can define

$$P_{-1}(\theta|S_{-1}) = \Pr\{\text{Test } S_{-1} \text{ accepts } H_{-1} \text{ when } \theta \text{ is true}\},$$

for probabilities of error given by (3.2), and similarly we define

$$P_1(\theta|S_1) = \Pr\{\text{Test } S_1 \text{ accepts } H_1 \text{ when } \theta \text{ is true}\},$$

with the corresponding probabilities of error. For the combined test $S$, write the probability of accepting $H_i$ when $\theta$ is true as $P_i(\theta|S)$, for $i = -1, 0, 1$. Clearly

$$P_{-1}(\theta|S) = P_{-1}(\theta|S_{-1})$$

since this is the probability of a path starting at the origin reaching the boundary $CM$ without first touching $BL$ (see Fig. 3.2). Similarly we have

$$P_1(\theta|S) = P_1(\theta|S_1).$$

Because the combined procedure $S$ terminates with probability one, then

$$\begin{aligned} P_0(\theta|S) &= 1 - P_1(\theta|S) - P_{-1}(\theta|S) \\ &= 1 - P_1(\theta|S_1) - P_{-1}(\theta|S_{-1}). \end{aligned} \tag{3.3}$$

This equation is exact, but the probabilities $P_1(\theta|S_1)$ and $P_{-1}(\theta|S_{-1})$ are known only approximately, by Theorem 2.2. In particular, if $\theta = \frac{1}{2}$,

$$P_0(\tfrac{1}{2}|S) \simeq 1 - \alpha/2 - \alpha/2 = 1 - \alpha,$$

and the wrong terminal decision is reached by $S$ with probability approximately $\alpha$, where the approximations involved are merely those of setting up the individual tests $S_{-1}$ and $S_1$.

Sobel and Wald (1949) obtained bounds for the ASN of the combined test $S$, for various ranges of $\theta$. They point out that the sample size for $S$ must be at least that required for either $S_{-1}$ or $S_1$ separately. Thus

$$E(n|\theta, S) \geqslant \text{Max}\{E(n|\theta, S_{-1}), E(n|\theta, S_1)\}, \tag{3.4}$$

where approximations to $E(n|\theta,S_i)$ are given in Theorem 2.3. For $\theta < \theta_1$ or $\theta > 1-\theta_1$, the lower bound (3.4) will be close.

To obtain some upper bounds proceed as follows. Consider test $S_{-1}$, then when $\theta < \theta_1$ the ASN is less than that required by a test which terminates only with acceptance of $H_{-1}$, and this latter quantity is approximately

$$\log(A)/E(z|\theta)$$

in the notation of Theorem 2.3. By proceeding in this way, Sobel and Wald obtained upper and lower bounds for the ASN for various ranges of $\theta$ (see their paper for details).

Sobel and Wald (1949) actually discuss this three-decision problem in terms of a test of a mean $\theta$ of a normal distribution with known variance $\sigma^2$. They suppose that if $\theta < \theta_1$, an hypothesis $H_{-1}$ is preferred; if $\theta_2 < \theta < \theta_3$ an hypothesis $H_0$ is preferred; and if $\theta > \theta_4$ an hypothesis $H_1$ is preferred, where $\theta_1 < \theta_2 < \theta_3 < \theta_4$, and the intermediate regions $(\theta_1, \theta_2)$ and $(\theta_3, \theta_4)$ are indifference regions. The discussion above is given to present the methods involved, and is based on the binomial, rather than the normal distribution. The presentation is simplified by putting $\theta_2 = \theta_3$.

For extreme values of $\theta$, when one of the SPRT's dominates, Theorem 2.4 will apply approximately. Presumably it is mainly this property to which Sobel and Wald refer when they say that they believe 'the suggested sequential procedure is not far from being optimum'. For intermediate values of $\theta$, near where $H_0$ is true, the termination rule used eliminates the possibility of any optimal property, although it is possible that the loss in efficiency is not great.

The main advantages of this method of testing three hypotheses lie in that it is a simple method of constructing a test for which there is a good approximation to the OC-curve, and some approximate information on the ASN-function.

### 3.4. Armitage's Method

So far we have obtained a sequential test of three hypotheses by using two of the three possible SPRT's. Armitage (1950) suggested that, in effect, all three SPRT's be used. We shall discuss his suggestion in terms of Ex. 3.1, with hypotheses (3.1), and obtain some boundaries in the lattice diagram, Fig. 3.1.

Suppose we operate all three SPRT's until there is a set of results such that for some $H_i$, $H_i$ is preferred to $H_j$ for both $H_j$, $j \neq i$, on the basis of the individual SPRT's. This leads to two important differences in the boundaries from those shown in Fig. 3.1. Firstly the difficulty of boundaries beneath $BC$ is avoided by performing an SPRT of $H_{-1}$ versus $H_1$ in this region. Secondly, sampling terminates only when a decision is reached simultaneously by both of the relevant SPRT, so that the path plotted in Fig. 3.2 would not be terminated as shown, but $H_0$ would be accepted only when the area $KAL$ is reached. These points can be seen easily in Fig. 3.3.

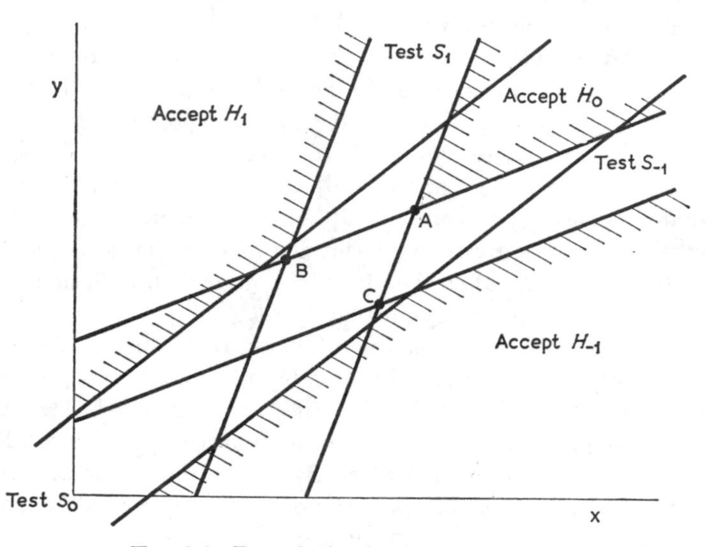


FIG. 3.3. Boundaries for Armitage's test.

Armitage defends this suggestion by analogy with linear discriminant function analysis, and undoubtedly these boundaries have the greater intuitive appeal. Unfortunately, our knowledge of the properties of such boundaries is very scanty. We proceed as follows. Consider again the yarn-testing example, Ex. 3.1. Let the number of breaks in yarn $A$ and yarn $B$ after $r$ trials be $y_r$ and $x_r$ respectively, and denote the probability of the results $(y_r, x_r)$ given $H_i$ by $p(y_r, x_r | H_i)$. Then the SPRT of $H_i$ versus $H_j$ can be written

$$\frac{1}{A_{ji}} < \frac{p(y_r, x_r | H_i)}{p(y_r, x_r | H_j)} < A_{ij} \qquad \text{where the } A_{ij} \text{ are constants.} \qquad (3.5)$$

We now follow the reasoning in Theorem 2.1 and add up inequalities (3.5), over all results leading to $H_i$ being accepted. (Notice that for $H_i$ to be accepted, the SPRT's for both $H_i$ versus $H_j$ and $H_i$ versus $H_k$ must give a preference for $H_i$.) This procedure yields

$$\pi_{ii}/\pi_{ij} > A_{ij}$$

and

$$\pi_{ii}/\pi_{ik} > A_{ik}$$

where $\pi_{ij}$ is the probability that the procedure terminates with acceptance of $H_i$ when $H_j$ is true. Now $\pi_{ii} < 1$, so that we have an inequality for $\pi_{ij}$,

$$\pi_{ij} < 1/A_{ij}.$$

Now, since the suggested test is a series of SPRT's, it must eventually terminate, and therefore

$$\pi_{-1,i}+\pi_{0,i}+\pi_{1,i} = 1.$$

Therefore, for example

$$\begin{aligned}\pi_{0,i} &= 1-\pi_{-1,i}-\pi_{1,i}\\ &\geqslant 1-1/A_{-1,i}-1/A_{1,i}.\end{aligned}$$

If we put $A_{ij} = (1-\alpha)/\alpha$, all $i,j$, then

$$\pi_{ii} > 1-2\alpha/(1-\alpha), \tag{3.6}$$

for $i = -1, 0, 1$. However, if $BC$ is not in the positive quadrant there is only a small difference between these boundaries and those given by the earlier theory (this difference being due to the paths which cross both $AB$ and $AC$, Fig. 3.2). Thus approximately, from (3.3)

$$\pi_{-1,-1} \simeq 1-\alpha \simeq \pi_{1,1}$$

but from (3.6),

$$\pi_{1,1} \geqslant 1-2\alpha/(1-\alpha),$$

and

$$\pi_{-1,-1} \geqslant 1-2\alpha/(1-\alpha).$$

This indicates that two of the three inequalities (3.6) may be unnecessarily wide.

We have almost no knowledge of the ASN-function of this procedure. If $BC$ is not in the positive quadrant, the lower bound to the ASN-function obtained by Sobel and Wald will hold here, and some of the upper bounds will also be true.

We may therefore conclude this discussion as follows. If $BC$ is not in the positive quadrant, there is often very little difference between the two types of boundaries, and Armitage (1950) would seem the more natural to use, if more of the properties of the boundaries were known. If $BC$ is in the positive quadrant then we can use the Armitage (1950) method, but we have only inequalities on the OC-function which may be unnecessarily wide, and there is no information about the ASN-function.

### 3.5 Billard and Vagholkar's Test

Billard and Vagholkar (1969), based on an idea due to D. R. Cox, have put forward another procedure for making a two-sided test of a normal mean, which they believe may be close to being optimum for a certain range of the parameter under test.

Suppose random variables $x_i$, for $i=1, 2, \ldots$, are independently and normally distributed with a known variance $\sigma^2$, and it is required to test the null hypothesis $H_0$: $\theta = 0$, where $\theta$ is the unknown mean, against the two-sided alternative $H_1$: $\theta = \pm\theta_1$, then the test follows the plan shown in Fig. 3.4. In this procedure, no decision is made until a sample size $n_0$ is reached. If at $n=n_0$ the boundaries {A, BC or DM are reached, sampling stops and the appropriate decisions are made. If at $n=n_0$ we hit the region AB or CD, sampling continues to distinguish between the decisions $\theta=0$ and $\theta=\theta_1$ in the first case, and between $\theta=0$ and $\theta=-\theta_1$ in the second. Now sampling in each of these continuation regions can be reduced to the consideration of a random walk operating between parallel absorbing boundaries, and formulae for the OC-curve and ASN function in this case are well known. Therefore by using these formulae, and integrating over the sampling distribution of $x$ at $n=n_0$, it is possible to obtain formulae for the OC-curve and ASN function of the procedure.

There remains the question of fixing the parameters ($n_0$, a, b, $\psi$) which characterize the procedure, and clearly a number of approaches could be adopted here. Billard and Vagholkar had the restrictions

$$\Pr\{\text{Decision for } H_0 | H_0 \text{ true}\} \geq 1-\alpha$$

and

$$\Pr\{\text{Decision for } H_0 | \theta = \pm\theta_1\} \leq \beta,$$

and then determined the set of parameters which minimized the ASN at a specified value of $\theta$. The minimization was carried out using a computer optimization algorithm, and three sets of plans were produced, with the ASN minimized at $\theta=0$, $\theta=\theta_1$, and $\theta=(\theta_0+\theta_1)/2$. The OC-curves and ASN functions of the three sets of plans are in fact very similar, and this is one evidence quoted by the authors that the plans are close to optimal.

Table 3.1 shows the OC-curve and ASN function of Billard and Vagholkar's plan determined theoretically, and by empirical sampling, and also results determined empirically on an equivalent Sobel and Wald plan. The results indicate that this new procedure may be close to optimal for $|\theta| \leq \theta_1$. (See further results in the source paper).

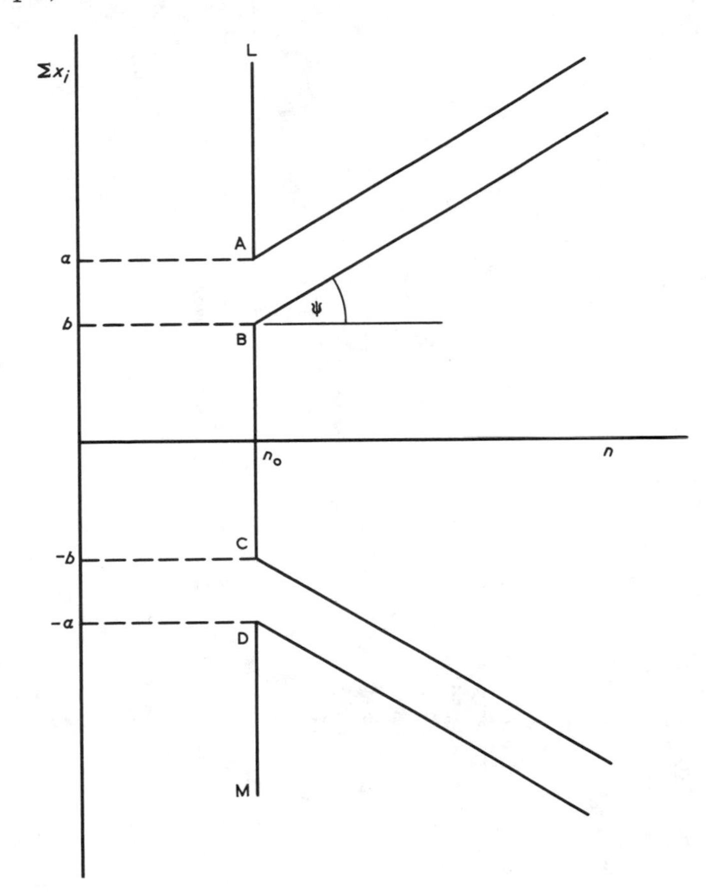


FIG. 3.4. Billard and Vagholkar's sequential plan.

TABLE 3.1 A comparison of Billard and Vagholkar's and Sobel and Wald's sequential plans

| | *OC-curve* | | | *ASN -function* | | |
|---|---|---|---|---|---|---|
| $\theta$ | Theoretical | Empirical | S. & W. | Theoretical | Empirical | S. & W. |
| 0 | 0·950 | 0·966 | 0·972 | 8·06 | 9·75 | 10·63 |
| 0·2 | 0·893 | 0·900 | 0·930 | 8·73 | 10·13 | 11·84 |
| 0·4 | 0·693 | 0·697 | 0·725 | 10·02 | 13·27 | 14·67 |
| 0·6 | 0·391 | 0·359 | 0·347 | 10·15 | 13·57 | 14·76 |
| 0·8 | 0·158 | 0·128 | 0·100 | 8·80 | 11·34 | 11·64 |
| 1·0 | 0·050 | 0·027 | 0·022 | 7·24 | 8·83 | 8·30 |
| 1·2 | 0·013 | 0·007 | 0·007 | 6·18 | 7·74 | 6·20 |

Note: The parameters for Billard and Vagholkar's plan are such that the ASN is minimized at the value $\theta=0$. The properties of Sobel and Wald's plan were determined empirically.

### 3.6. Some Special Methods of Treating Three Hypotheses

In some special cases a problem involving a test of three hypotheses can be put as a test of only two hypotheses, so that the standard SPRT can be used.

Armitage (1947) suggests a test that the mean of a normal distribution is $\mu_0$ against two-sided alternatives as follows. Consider the sign of the difference between each observation and $\mu_0$, and take observations in pairs. Denote the combinations $-+$ or $+-$ by $x$, and $++$ or $--$ by $y$, then under the null hypothesis the probability of $x$ is $\frac{1}{2}$, whereas under either of the (two-sided) alternatives it is less than a half. In fact, we have

$$\Pr(x) = 1-\{\Pr(\text{obs. positive})\}^2-\{\Pr(\text{obs. negative})\}^2.$$

If we now test the hypothesis that $\Pr(x) = \frac{1}{2}$ against some suitable alternative $\Pr(x) < \frac{1}{2}$, we have a two-sided test of the null hypothesis that the mean of the observations is $\mu_0$, by using just one SPRT. There will be some loss in efficiency in this procedure, owing to the pairing of observations, and because only the sign of the difference of each pair is used.

Another test can be obtained as follows. Take observations one at a time, and consider only the sign of the difference from $\mu_0$. Let $r$ denote the number of positive differences in a sequence of $n$ (non-zero) differences, then the probability distribution of $r$ is

$$\Pr(r|n) = {}^nC_r\theta^r(1-\theta)^{n-r},$$

for $r = 1, 2, \ldots, n$, where $\theta$ is the probability of a positive difference. Put $s = |r-\frac{1}{2}n|$, the absolute difference between $r$ and the expectation of $r$ under the null hypothesis, then the probability distribution of $s$ is

$$\Pr\{s|n\} = \tfrac{1}{2}\,{}^nC_{s+\frac{1}{2}n}\{\theta(1-\theta)\}^{\frac{1}{2}n}\left\{\left(\frac{\theta}{1-\theta}\right)^s+\left(\frac{1-\theta}{\theta}\right)^s\right\} \tag{3.7}$$

where the upper limit of $s$ is $(\frac{1}{2}n)$ and the lower limit is 0 or $\frac{1}{2}$ depending on whether $n$ is even or odd. The probability (3.7), as a function of $\theta$, is symmetrical about $\theta = \frac{1}{2}$, when (3.7) reduces to

$$\Pr\{s|n\} = {}^nC_{s+\frac{1}{2}n}(\tfrac{1}{2})^n. \tag{3.8}$$

We therefore form an SPRT using the distribution (3.7) to test $H_0:\theta = \frac{1}{2}$ against $H_1:\theta = \theta_1$, with probabilities of error $\alpha$ and $\beta$, as defined in Chapter 2. This leads to the test specified by

$$B < 2^n\{\theta(1-\theta)\}^{\frac{1}{2}n}\left\{\left(\frac{\theta}{1-\theta}\right)^s + \left(\frac{1-\theta}{\theta}\right)^s\right\} < A,$$

following the standard SPRT theory. This test will be discussed further in § 4.3.

For other special methods for two-sided tests of a null hypothesis see Gilchrist (1961) and Hajnal (1961). (Also see P3.3 and Ex. 4.1.)

### 3.7. Application of the Methods

In the references, a number of applications of the methods discussed in this Chapter are listed. Apart from some sequential medical trials, applications seem to have been rather rare, and, as in Chapter 2, some of the applications are unsatisfactory from various points of view.

Anderson's yarn testing problem (Ex. 3.1) is a direct application of Armitage's (1950) method. It would be difficult to take lengths of yarn randomly, but the theory probably holds reasonably well provided the testing of adjacent lengths is avoided, for example, by discarding a random length between lengths selected.

The insect inspection problem (Ex. 3.2) involved use of the negative binomial distribution (formula (2.20)), and a value of $k$ was carefully estimated from data. Sobel and Wald's (1949) formulation was used, except that no terminal decision was reached for paths not reaching the shaded areas of Fig. 3.2. The only properties worked out are for the separate SPRT's, which are not really relevant, and it is not known what effect errors or variations in $k$ would have. The field trial against the standard procedure referred to above appears to have been rather a small one.

Both Ex. 3.1 and Ex. 3.2 are very suitable for sequential methods, since a wide variety of conditions would be encountered in practice, and the sequential plans suggested would no doubt save observations on average, in comparison with a fixed sample size test.

All of the examples reveal the arbitrary nature of the hypotheses to formulate the test. I suspect that what actually happens in some cases (not necessarily those above!) is a sort of iterative procedure. Various combinations of hypotheses and risks are tried until a plan results which appears reasonable. This procedure can be defended on the grounds that the experimenter is balancing the cost of observations against the losses due to wrong decisions, in an intuitive way.

**Problems 3**

1. Suppose that observations $x_i$ are known to be independently and normally distributed with unit variance, but unknown mean $\mu$. Set up Sobel and Wald's sequential test of $H_0: \mu = 0$ against $H_1: \mu = \pm 0{\cdot}2$, with probabilities of error 1%. Calculate the bounds on the ASN-function given by Sobel and Wald (1949). For comparison, obtain a fixed sample-size procedure for testing the same hypotheses. Since the fixed sample-size plan and the sequential plan will clearly have different OC-curves, on what basis do you consider the two plans to be equivalent?

2. Schneiderman and Armitage (1962a, p. 53, Table 7) present the results of some empirical sampling trials on Sobel-and-Wald type two-sided tests of a normal mean, with known variance. Examine how the observed ASN- and OC-curves compare with Sobel and Wald's theory.

3. Suppose observations $x_i$ are known to be independently and normally distributed with unit variance, and it is required to test hypotheses about the unknown mean $\mu$, $H_0: \mu = 0$, against $H_1: \mu = \pm \mu_1$. If we consider the quantities $x_1^2, x_2^2, \ldots$, then under $H_0$ these are distributed independently $\chi_1^2$, while under $H_1$ (for $\mu = +\mu_1$ or $\mu = -\mu_1$), the quantities are independently distributed as non-central $\chi^2$, with non-centrality parameter $\mu_1^2$. Obtain an SPRT for testing $\mu_0$ against $\mu_1$ by proceeding in this way, and examine some properties of the test.

Is this test fully efficient, and if not, why? [Hint: Consider the efficiency of the fixed sample size test obtained on the lines indicated above.]

CHAPTER 4

# Extensions to the SPRT

## 4.1. The Problem of Composite Hypotheses

Suppose we can observe random variables $x_i$ which are known to be normally distributed, but with unknown mean and variance, then it frequently arises that we wish to test hypotheses about the location of the distribution, the test to be valid regardless of the value of the unknown variance. In discussing this problem Wald (1947) suggests that it might often be reasonable to impose the following structure. For all $\mu$ satisfying $|(\mu-\mu_0)/\sigma| < \delta_0$, where $\delta_0$ is small, it is preferred to accept an hypothesis $H_0$, and for all $\mu$ satisfying $|(\mu-\mu_0)/\sigma| > \delta_1$, it is preferred to accept an alternative hypothesis $H_1$, while there is an intermediate region of indifference. This formulation is equivalent to testing hypotheses about the proportion of the distribution which is above or below $\mu_0$. Even with this structure, the SPRT theory given in the previous chapters does not lead to a test, for the likelihood ratio depends on the unknown $\sigma$. For example, for a one-sided test of $H_0 : \mu = \mu_0$ against $H_1 : \mu = \mu_0 + \delta\sigma$ (so that $H_1$ is a composite hypothesis) the log-likelihood ratio is

$$n(\bar{x}-\mu_0)\delta/\sigma - n\delta^2/2$$

which depends on $\sigma$. (If bounds can be assumed on $\sigma$ some rough conservative procedure can be worked out, but this would usually be very inefficient.) A satisfactory test would have to have maximum probabilities of error $\alpha$ and $\beta$ in the respective regions in which $H_0$ and $H_1$ are preferred: it seems clear from invariance considerations that a satisfactory test exists, since the power of the $t$-test depends only on $(\mu-\mu_0)/\sigma$ (and the degrees of freedom). However, some extension is needed to the SPRT theory to enable us to cope with situations in which either or both the hypotheses are composite.

Before proceeding with the extensions, it will be helpful to consider very briefly an allied problem. Suppose we have observations distributed according to the negative binomial distribution (2.20), with

$k$ and $\theta$ both unknown, and suppose that we wish to test hypotheses about the mean of the distribution, the tests to be valid regardless of the unknown $k$. (This problem arises in both Ex. 2.2 and Ex. 3.2.) The precise formulation of the hypotheses to be tested could be done in a number of ways, analogous to that for the normal distribution above; we specify two regions of the space $(k, k\theta)$, in which two hypotheses $H_0$ and $H_1$ respectively will be preferred, and the maximum probabilities of error to be $\alpha$ and $\beta$ in the two regions. If the full range of $k$ is allowed, it is not clear whether a satisfactory test exists, in contrast with the normal distribution case, where the existence of a test is not in doubt.

In this chapter we discuss two general methods which sometimes lead to a sequential test when the hypotheses are composite.

## 4.2. Wald's Theory of Weight Functions

Assume that random variables $x_i$ are observed which are independently distributed with a probability density function $f(x, \theta)$, where $\theta = \{\theta_1, \ldots, \theta_k\}$. Suppose the space of $\theta$ is divided into three regions, $\omega_0$ in which an hypothesis $H_0$ is preferred, $\omega_1$ in which an hypothesis $H_1$ is preferred, and the remainder an indifference region. We require

$$\begin{aligned} &\text{Max}\{\text{probability of accepting } H_1 \text{ for } \theta \epsilon \omega_0\} \leqslant \alpha \\ &\text{Max}\{\text{probability of accepting } H_0 \text{ for } \theta \epsilon \omega_1\} \leqslant \beta, \end{aligned} \tag{4.1}$$

and the test must terminate with probability unity.

Suppose we have two prior distributions for $\theta$, one $W_0(\theta)$ which is non-zero only within $\omega_0$, and the other $W_1(\theta)$ which is non-zero only within $\omega_1$, so that

$$\int_{\omega_0} W_0(\theta)\, d\theta = \int_{\omega_1} W_1(\theta)\, d\theta = 1. \tag{4.2}$$

Consider the two simple hypotheses, $H_i'$, that the probability of a set of observations $x_1, x_2, \ldots, x_n$ is for each fixed $n$,

$$\int_{\omega_i} f(x_1, \theta) f(x_2, \theta) \ldots f(x_n, \theta)\, W_i(\theta)\, d\theta, \tag{4.3}$$

for $i = 0, 1$. These hypotheses should be clearly distinguished from the simple hypothesis $H_i''$, that the probability density of each observation $x$ is

$$\int_{\omega_i} f(x, \theta)\, W_i(\theta)\, d\theta. \tag{4.4}$$

Hypotheses (4.3) state that the sampling for $\theta$ is done once only for a series of observations $x_1, \ldots, x_n$, whereas hypotheses (4.4) state that sampling for $\theta$ is done before each observation. Wald uses the form (4.3), and we note that the observations are not (usually) independent under these hypotheses. However, if the reader examines the proof of Theorem 2.1 closely, he will see that it nowhere assumes the observations to be independent. Theorem 2.1 can therefore be used to construct an SPRT for hypotheses (4.3), using probabilities of error $(\alpha',\beta')$, and the theorem implies that provided the test terminates with probability one

$$\int_{\omega_0} W_0(\theta)\,\alpha(\theta)\,\mathrm{d}\theta \simeq \alpha'$$

$$\int_{\omega_1} W_1(\theta)\,\beta(\theta)\,\mathrm{d}\theta \simeq \beta', \tag{4.5}$$

where $\alpha(\theta)$, $\beta(\theta)$ are the probabilities of error for observations distributed with a density $f(x,\theta)$.

The original requirements on the test were (4.1) which we write in the form

$$\alpha(\theta) \leqslant \alpha \qquad \text{for all } \theta\epsilon\omega_0,$$

and

$$\beta(\theta) \leqslant \beta \qquad \text{for all } \theta\epsilon\omega_1, \tag{4.6}$$

so that we must have $\alpha' \leqslant \alpha$, $\beta' \leqslant \beta$. Now for any given weight functions,

$$\operatorname*{Max}_{\theta\epsilon\omega_0} \alpha(\theta) \text{ and } \operatorname*{Max}_{\theta\epsilon\omega_1} \beta(\theta)$$

are functions of $\alpha'$, $\beta'$, the probabilities of error used in setting up the test of hypotheses (4.3). In principle, therefore, $\alpha'$ and $\beta'$ can be chosen so that $\alpha$ and $\beta$ take on their desired values.

If there is some basis for assuming particular prior distributions $W_0(\theta)$, $W_1(\theta)$, and yet a test is required satisfying (4.1), this theory provides one possible approach. (In this circumstance a better method of approach would be by the methods of Chapter 7.) However, usually there will be little or no basis for particular prior distributions, and they must be regarded as convenient mathematical devices – weight functions – used to reduce a difficult problem (involving composite hypotheses), to one already treated (tests of simple hypotheses). In principle, the choice of the weight functions could be made to satisfy some kind of optimality criterion, but unfortunately there is no general method available for choosing weight functions with suitable properties. Wald proceeds by searching for the weight functions for which the maxima of $\alpha(\theta)$ and $\beta(\theta)$ take on their minimum values. Wald then seeks to adjust the chosen values $\alpha'$ and $\beta'$ of (4.5) so that these maxima take on their desired values.

This approach has not been felt to be entirely satisfactory (see Barnard, 1947), and I have not found any major application of it other than those given by Wald himself. We shall not discuss Wald's method by which optimum weight functions can sometimes be determined, but instead we give three examples in some detail.

## 4.3. Some Applications of Wald's Theory of Weight Functions

*Ex. 4.1.* Suppose the observations $x_i$ are 1 or 0 with probability $\theta$ and $(1-\theta)$ respectively, and suppose we wish to test $H_0: \theta = \frac{1}{2}$ against two-sided alternatives, the region of preference for $H_1$ being $|\theta - \frac{1}{2}| > \delta$. Since the hypothesis $H_0$ is simple, we can choose $W_0(\theta)$

$$W_0(\tfrac{1}{2}) = 1$$
$$W_0(\theta) = 0, \qquad \text{for } \theta \neq \tfrac{1}{2}.$$

Choose the weight function $W_1(\theta)$ as

$$W_1(\theta_1) = W_1(1-\theta_1) = \tfrac{1}{2}$$
$$W_1(\theta) = 0, \quad \text{otherwise,}$$

where $\theta_1 = \frac{1}{2} + \delta$. Then, following (4.3), the modified form of the likelihood of results $(n, r)$ given $H_1$ is

$$\tfrac{1}{2}{}^nC_r\{\theta_1^r(1-\theta_1)^{n-r} + \theta_1^{n-r}(1-\theta_1)^r\}$$

where $r$ is the number of positive observations in a total of $n$. The SPRT is therefore specified by the inequalities

$$B < 2^{n-1}\{\theta_1^r(1-\theta_1)^{n-r} + \theta_1^{n-r}(1-\theta_1)^r\} < A. \tag{4.7}$$

For this test we have approximately

$$\text{Max Pr}\{\text{accept } H_0 \text{ when } \theta < 1-\theta_1 \quad \text{or} \quad \theta > \theta_1\} \simeq \beta$$
$$\text{Pr}\{\text{accept } H_1 \text{ when } \theta = \tfrac{1}{2}\} \simeq \alpha,$$

when $A = (1-\beta)/\alpha$ and $B = \beta/(1-\alpha)$, and where $H_1$ denotes the (original) alternative hypothesis.

Denote $y = r$ and $x = n - r$, then the boundaries for the test specified by (4.7) have a shape such as that shown in Fig. 4.1. In this figure, the dotted lines correspond to Sobel and Wald's test (§ 3.2) for this problem, which combine the tests $\{H_0:\theta=\frac{1}{2},\ H_1:\theta=\theta_1\}$ with probabilities of error $(\alpha/2,\beta)$ and $\{H_0:\theta=\frac{1}{2}\ H_1:\theta=1-\theta_1\}$ with the same probabilities of error.

Both channels of this test are open, and the upper and lower ends of the middle boundary tend to straight lines corresponding to part of the boundaries for tests with probabilities of error $(\alpha,2\beta)$, of $H_0:\theta=\frac{1}{2}$ against $H_1:\theta=\theta_1$ and $H_1:\theta=1-\theta_1$.

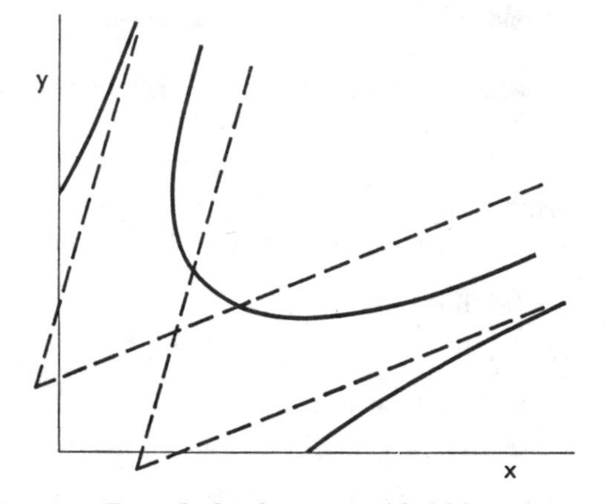


FIG. 4.1. Boundaries for a two-sided binomial test.

No detailed properties of this test are known, and no comparisons with the methods of Chapter 3 have been made. However, it appears from Fig. 4.1 that the sample size distribution of this test should be rather less variable than for Sobel and Wald's or Armitage's methods. Further, the probabilities of error $(\alpha,\beta)$, for test (4.7) should be set with reasonable accuracy, so that the test combines some of the better features of the methods of § 3.3 and § 3.4. For many problems the weight function approach is probably the best way of forming a two-sided test of a null hypothesis by means of the SPRT. See problem P4.6.

Wald (1947) proposed a test analogous to (4.7) for a two-sided test of a normal mean (variance assumed known).

*Ex. 4.2. Wald's sequential* t*-test.* In § 4.1 we formulated a problem of testing a normal mean when $\sigma$ is unknown. We specify three regions, in the space $(\theta,\sigma)$, two in which $H_0$ and $H_1$ respectively are preferred,

$$H_1: |\theta-\theta_0|/\sigma > \delta_1,$$
$$H_0: \theta = \theta_0$$

while the remainder is an indifference region. For this problem we need weight functions for both $\theta$ and $\sigma$ and Wald suggests the weight function

$$\begin{aligned} \phi(\sigma) &= \frac{1}{c}, \qquad 0 \leqslant \sigma \leqslant c \\ &= 0, \qquad \text{otherwise,} \end{aligned} \tag{4.8}$$

for both $W_0$ and $W_1$. For $\theta$ we choose

$$\begin{aligned} W_0: \psi_0 &= 1 \\ \psi_0 &= 0 \text{ for all other } \theta. \\ W_1: \psi_1 &= \tfrac{1}{2} \qquad \text{when } \theta = \theta_0 \pm \delta_1\sigma \\ &= 0, \qquad \text{otherwise.} \end{aligned}$$

Thus the modified likelihoods of observations given $H_0$ and $H_1$ are respectively

$$\int_0^c \frac{1}{\sigma^n} \exp\left\{-\frac{1}{2\sigma^2}\Sigma\,(x_i-\theta_0)^2\right\} d\sigma / c(2\pi)^{\frac{1}{2}n},$$

and

$$\left[\int_0^c \frac{1}{\sigma^n} \exp\left\{-\frac{1}{2\sigma^2}\Sigma\,(x_i-\theta_0-\delta\sigma)^2\right\} d\sigma + \int_0^c \frac{1}{\sigma^n} \exp\left\{-\frac{1}{2\sigma^2}\Sigma\,(x_i-\theta_0+\delta\sigma)^2\right\} d\sigma\right] \bigg/ 2c(2\pi)^{\frac{1}{2}n} \tag{4.9}$$

If we take the likelihood ratio and proceed to the limit $c \to \infty$, we have the ratio

$$\frac{\tfrac{1}{2}\left[\int_0^\infty \frac{1}{\sigma^n} \exp\left\{-\frac{1}{2\sigma^2}\Sigma\,(x_i-\theta_0-\delta\sigma)^2\right\} d\sigma + \int_0^\infty \frac{1}{\sigma_n} \exp\left\{-\frac{1}{2\sigma^2}\Sigma\,(x_i-\theta_0+\delta\sigma)^2\right\} d\sigma\right]}{\int^\infty \frac{1}{\sigma_n} \exp\left\{-\frac{1}{2\sigma^2}\Sigma\,(x_i-\theta_0)^2\right\} d\sigma} \tag{4.10}$$

Wald shows that this likelihood ratio is a single-valued, strictly increasing function of $|(\bar{x}-\theta_0)/\sqrt{\sum(x_i-\bar{x})^2}|$, and from this it follows that $\alpha(\theta,\sigma)$ is constant in $\omega_0$, and $\beta(\theta,\sigma)$ is constant whenever

$$\theta = \theta_0 \pm \delta\sigma$$

which is the only region in which $W_1(\theta,\sigma)$ is non-zero. Thus the ratio (4.10) can be computed and used in an SPRT with strength $\alpha,\beta$, and we expect the actual hypotheses $H_0$ and $H_1$ to be tested with approximately these probabilities of error.

*Ex. 4.3. The sequential $\chi^2$-test.* Suppose that $x_i$ are independent normally distributed random variables, and the mean and variance are both unknown. We wish to test the hypothesis $H_0: \sigma=\sigma_0$, against $H_1: \sigma=\sigma_1>\sigma_0$.

In this example the unknown mean $\theta$ is a nuisance parameter, and we choose the weight function

$$\begin{aligned}\phi(\theta) &= \frac{1}{2c}, \qquad -c \leqslant \theta \leqslant +c\\ &= 0 \qquad \text{otherwise.}\end{aligned}$$

The modified likelihood for $H_i'$ is

$$\frac{\exp\left(-\dfrac{(n-1)s^2}{2\sigma_i^2}\right)\displaystyle\int_{-c}^{c}\frac{\sqrt{n}}{\sqrt{(2\pi)}\,\sigma_i}\exp\left\{-\frac{n}{2\sigma_i^2}(\bar{x}-\theta)^2\right\}d\theta}{\{2c\sqrt{n}(2\pi)^{(n-1)/2}\sigma_i^{n-1}\}} \tag{4.11}$$

for $i=0$, 1, where $s^2$ is the sample variance. By taking the ratio $H_1'/H_0'$, and proceeding to the limit $c\to\infty$, we have the ratio,

$$\left(\frac{\sigma_0}{\sigma_1}\right)^{n-1}\exp\left\{+(n-1)\frac{s^2}{2}\left(\frac{1}{\sigma_0^2}-\frac{1}{\sigma_1^2}\right)\right\} \tag{4.12}$$

which we can use in an SPRT. It is easily seen that in this example the actual value of $\theta$ makes no difference to the probabilities of error, so that an SPRT for $H_0'$ vs. $H_1'$ with strength $\alpha,\beta$, also has strength approximately $\alpha,\beta$ for testing $H_0$ against $H_1$.

This particular example is of further interest in that the ratio (4.12) has identical properties for a sequence obtained on $(n-1)$ observations with known mean, say zero, and variance equal to the variance of the $x_i$. This is easily established using the Helmert transformation

$$z_1 = \frac{1}{\sqrt{n}}\{x_1+\ldots+x_n\}$$

$$z_j = \frac{1}{\sqrt{\{j(j-1)\}}}\{x_1+\ldots+x_{j-1}-(j-1)\,x_j\} \qquad \text{for } j = 2,\ldots,n.$$

Whence

$$s^2 = \left\{\sum_{2}^{n} z_j^2/(n-1)\right\},$$

and the result follows. Therefore the properties of this sequential test can be worked out by the theory of Chapter 2 in this particular case.

*Discussion*

Wald limits himself to considering (modified) hypotheses of the form (4.3) because by using relation (4.5) tests can be constructed with approximately desired limits on $\alpha$ and $\beta$. Except in special cases (as in Ex. 4.3), the modified hypotheses result in the observations not being independent, and the Wald–Wolfowitz theorem (Theorem 2.4) does not hold, nor can the OC-curves and ASN-functions be evaluated by the methods given in Chapter 2.

This method will yield an infinite set of sequential tests, each corresponding to a different set of weight functions. Wald's suggestion of choosing those giving the lowest maximum for $\alpha$ and $\beta$ cannot in general be achieved; even when this minimax property holds, it is not clear what properties the ASN-function has.

**4.4. Transformation of Observations**

The method suggested by Wald for circumventing the difficulty of forming sequential tests for composite hypotheses, is merely to form modified hypotheses by integrating out the nuisance parameter, etc. In this way he hopes a test can be found which will satisfy the original requirements.

A second way out of the difficulty is to consider a sequence formed by transforming the original observations, the transformation so chosen that the new sequence does not depend on nuisance parameters. For example, Armitage (1947) suggested a two-sided test of a normal mean, valid whether $\sigma$ is known or not, by considering the transformed sequence

$$\text{sign}(x_1-\mu), \quad \text{sign}(x_2-\mu), \quad \ldots, \tag{4.13}$$

see § 3.5.

For the test of a normal mean when $\sigma$ is unknown, mentioned in § 4.1, we could consider the transformed sequence $t_2, \ldots, t_n$, instead of the original observations $x_i$, where

$$t_k = \frac{\left(\sum_1^k x_i - k\mu_0\right)\sqrt{(k-1)}}{k\sqrt{\left[\sum_1^k (x_i-\bar{x})^2\right]}}. \tag{4.14}$$

The probability distribution of $t_i$ depends only on $\Delta = (\mu-\mu_0)/\sigma$ and since $(\bar{x}, s)$ are sufficient for $(\mu, \sigma)$, this transformed sequence should be capable of yielding a very efficient sequential test. The SPRT can be applied to this sequence directly, so that after $x_n$ we calculate the likelihood ratio

$$\frac{f\{t_2, \ldots, t_n | H_1\}}{f\{t_2, \ldots, t_n | H_0\}}, \tag{4.15}$$

where $f(t_2, \ldots, t_n | H_i)$ is the probability density function of $(t_2, \ldots, t_n)$ on $H_i$. However, these transformed sequences such as $t_i$, are not usually independent, and the likelihood ratio (4.15) could be very complicated to calculate. This method of forming a sequential test will usually only become practicable in cases where there is some great simplification.

For the test of a normal mean discussed above, it can be shown that the density of $(t_2, \ldots, t_n)$ factorizes into the density of $t_n$ multiplied by the density of $(t_2, \ldots, t_{n-1})$ conditional on $t_n$, which we write as

$$f\{t_2, \ldots, t_n | \Delta\} = f_1\{t_n | \Delta\} f_2\{t_2, \ldots, t_{n-1} | t_n\}, \tag{4.16}$$

where the second term is independent of $\Delta$. Thus (4.15) simplifies to the ratio of the likelihoods of $t_n$ only, under $H_1$ and $H_0$. Cox's theorem (1952a) is a fixed sample size theorem which states specific conditions under which the factorization (4.16) takes place. Although the theorem is given in connection with the SPRT there are wider implications, to which we shall refer later.

## 4.5. Cox's Theorem

*Theorem 4.1* Let $x_1, x_2, \ldots, x_n = \underline{x}$ be random variables whose probability density function (p.d.f.) depends on unknown parameters $\theta_1, \theta_2, \ldots, \theta_p$. The $x_i$ may themselves be vectors.

Suppose that

(i) $y_1, y_2, \ldots, y_p$ are a functionally independent jointly sufficient set of estimators for $\theta_1, \theta_2, \ldots, \theta_p$;
(ii) the distribution of $y_1$ involves $\theta_1$ but not $\theta_2, \ldots, \theta_p$;
(iii) $u_1, \ldots, u_m$ are functions of $\underline{x}$ functionally independent of each other, and of $y_1, y_2, \ldots, y_p$;
(iv) there exists a set $S$ of transformations of

$$\underline{x} = \{x_1, x_2, \ldots, x_n\} \quad \text{into} \quad \underline{x}' = \{x_1', x_2', \ldots, x_n')$$

such that

(a) $y_1, u_1, \ldots, u_m$ are all unchanged by transformations in $S$;
(b) the transformation of $y_2, \ldots, y_p$, into $y_2', \ldots, y_p'$, defined by each transformation in $S$ is (1:1);
(c) if $Y_2, \ldots, Y_p$ and $Y_2', \ldots, Y_p'$ are two sets of values of $y_2, \ldots, y_p$ each having non-zero probability density under at least one of the distributions of $\underline{x}$, then there exists a transformation in $S$ such that if $y_2 = Y_2, \ldots, y_p = Y_p$, then $y_2' = Y_2', \ldots, y_p' = Y_p'$.

Then the joint p.d.f. of $y_1, u_1, \ldots, u_m$ factorizes into

$$g(y_1|\theta_1)\, l(u_1, \ldots, u_m|y_1) \tag{4.17}$$

where $g$ is the p.d.f. of $y_1$ and $l$ does not involve $\theta$.

A full and detailed proof of Cox's theorem is given by Cox (1952a), to which we refer the reader. It is important to understand clearly the object of the theorem before setting out on the proof. We have a set of independent random variables $\{x_1, \ldots, x_n\}$, and we are interested in making inferences about a parameter $\theta_1$. The p.d.f. of $\{x_1, \ldots, x_n\}$ depends on other parameters than $\theta_1$, and we consider a transformed set of variables $\{y_1, u_1, \ldots, u_m\}$, obtained from $\{x_1, \ldots, x_n\}$, such that the transformed set depends only on $\theta_1$. We want to obtain the p.d.f. of this transformed sequence, and to determine the conditions that it factorizes in the form given (4.17). For a more rigorous justification of Cox's method, derived from the principle of invariance, see Hall, Wijsman and Ghosh (1965, particularly pages 585–587).

*Ex. 4.4. Illustration of Cox's theorem for the sequential* t*-test.* Suppose $x$ are independent normally distributed random variables with

$E(x_i) = \delta\sigma$, $V(x_i) = \sigma^2$, and suppose we wish to test the null hypothesis $H_0$ that $\delta = 0$, against $H_1$ that $\delta \geqslant \delta_1 > 0$, and that $\sigma$ is unknown.

The discussion in § 4.4 establishes that we can consider the transformed sequence $t_2, t_3, \ldots, t_n$, where

$$t_k = \frac{\sum_{i=1}^{k} x_i}{k\sqrt{\left[\sum_{1}^{k} \frac{(x_i - \bar{x})^2}{k-1}\right]}} = \frac{\bar{x}}{s}$$

in standard notation, and we know further that the distribution of this sequence depends on $\delta$, and not on $\sigma^2$. Thus it will be legitimate to test $H_0$ against $H_1$ using (4.15).

To apply Cox's theorem, put $p = 2$, and

$$\theta_1 = \delta \qquad \theta_2 = \sigma$$
$$y_1 = t_n = \bar{x}/s, \qquad \text{as above}$$
$$y_2 = s = \sqrt{\left[\sum_{1}^{n} (x_i - \bar{x})^2/(n-1)\right]}.$$

Further, put $u_j = t_j$ for $j = 2, 3, \ldots, (n-1)$, and we notice that the $t_j$ are functionally independent (but not statistically independent) of each other and of $y_1$ and $y_2$. Further, the statistics $(t_n, s)$ are jointly sufficient for $(\delta, \sigma)$. We have now satisfied conditions (i), (ii), and (iii) of the theorem.

For the transformations $S$ use the set

$$\{z' = az; \quad a > 0\}$$

and $y_1, u_1, \ldots, u_m$ are unchanged by these transformations. Condition (c) of (iv) is satisfied by noticing that if $s^2$, $s'^2$ are any two values of sample variances, there is a value of the constant '$a$' for which $s'^2 = as^2$.

Thus by the theorem we can write

$$f\{t_2, t_3, \ldots, t_n | \delta\} = f_1\{t_n | \delta\} f_2\{t_2, t_3, \ldots, t_{n-1} | t_n\}$$

and the likelihood ratio (4.15) reduces to

$$\frac{f_1(t_n | \delta_1)}{f_1(t_n | \delta = 0)} \tag{4.18}$$

where $f_1(t_n|\delta)$ is the probability density function of $t_n$ given $\delta$, which is the non-central $t$-distribution, and can be written (see Rushton, 1952),

$$f_1\{t_n|\delta\} = \frac{\sqrt{n}\Gamma(n)\exp\{-\frac{1}{2}n(n-1)\,\delta^2/(n-1+nt_n^2)\}}{2^{\frac{1}{2}(n-2)}\Gamma\{\frac{1}{2}(n-1)\}\sqrt{\pi}\sqrt{(n-1)}}$$

$$\times\left(\frac{n-1}{n-1+nt_n^2}\right)^{\frac{1}{2}n} Hh_{n-1}(-\delta a_n)$$

where

$$a_n = nt_n/\sqrt{(n-1+nt_n^2)} = \sum_1^n x_i\bigg/\sqrt{\left[\sum_1^n x_i^2\right]}, \tag{4.19}$$

and where

$$Hh_n(x) = \int_0^\infty \frac{y^n}{n!}\mathrm{e}^{-\frac{1}{2}(y+x)^2}\,\mathrm{d}y.$$

The ratio (4.18) thus reduces to

$$\exp\left\{-\frac{\delta_1^2\,(n-a_n^2)}{2}\right\} Hh_{n-1}(-\delta_1 a_n)/Hh_{n-1}(0). \tag{4.20}$$

This is the likelihood ratio for use with (2.1) of § 2.2 to obtain an SPRT. The latter part of this proof (after application of Cox's theorem) is due to Rushton (1950), based on theory by Barnard (1952). Rushton also obtained an approximation to the log-likelihood ratio (4.20) suitable for practical use. We notice that the form (4.20) differs slightly from that obtained by Wald's weight function method, and this point will be discussed further in § 5.1.

The reader interested in following up references on sequential $t$-tests should notice that Rushton (1950) tests a null hypothesis $H_0\colon\delta = \delta_0$, and does not assume $\delta_0 = 0$. Rushton's likelihood ratio therefore results in a ratio of two non-central $t$-distributions rather than the ratio of non-central to central $t$-distributions obtained in (4.20).

*Ex. 4.5.* Suppose the response $y_i$ to an observation at a level $x_i$ is either 1, or 0, with probability of a positive result

$$\{1+\mathrm{e}^{-\theta_2(x-\theta_1)}\}^{-1}.$$

Sufficient statistics for $\theta_1$ and $\theta_2$ are $\sum y_i$, $\sum y_i x_i$. It may happen that we wish to test that $\theta_1$ has some specified value, when (the slope) $\theta_2$ is unknown. In spite of the existence of sufficient statistics, there is no way of combining them so that some function

$$\psi(\sum y_i, \sum y_i x_i)$$

has a distribution depending only on $\theta_1$. Cox's theorem cannot therefore be applied.

*Ex 4.6.* In Ex. 2.2 and Ex. 3.2 an SPRT for the mean of a negative binomial distribution was constructed by assuming a constant value of $k$. Suppose we wish to test the mean without any knowledge of $k$, then Cox's theorem cannot be applied, as no sufficient statistics exist.

## 4.6. Discussion

Cox's original paper contains a theorem which enables one to verify in a simple way that many tests constructed by the method indicated in § 4.5 above terminate with probability one. Unfortunately, Wald's method of obtaining approximations to the OC-curve and ASN-function usually breaks down, as with Wald's method of weight functions, because the series of transformed variables are dependent; thus very little can be said in general, about the properties of SPRT's obtained through Cox's theorem.

Cox (1952a) also gives an interesting example of an SPRT with non-independent observations, which has higher ASN's at both $H_0$ and $H_1$ than a test which is not of the SPRT type. The methods given so far in this chapter cannot be claimed to be much more than mathematical devices to obtain boundaries for which it is possible to set approximately certain probabilities of error.

In deciding which method to use in a particular case, it is useful to distinguish between two kinds of composite hypotheses:
(a) composite hypotheses involving ranges of the parameters of interest, as in the two-sided binomial test, Ex. 4.1;
(b) composite hypotheses involving nuisance parameters, as in the sequential $t$-test, Ex. 4.2, and Ex. 4.4.
Wald's method of weight functions is more suitable for case (a) than case (b), and Cox's theorem is more suitable for case (b). (But see P4.5.) Little is known to compare the two procedures, but for case (b) it seems more natural to use Cox's theorem, since if we have an hypothesis-testing problem in which there are unknown nuisance parameters, then we normally try to construct a test statistic having

a distribution not dependent on the nuisance parameter. This leads to considerations of functions of the observations which are invariant under a special group of transformations. We should not normally try to integrate out the nuisance parameters with some arbitrary weight functions.

### 4.7. Asymptotic Tests

If sufficient statistics do not exist, sequential tests cannot be constructed by Cox's theorem. However, Cox (1963) developed a method due to Bartlett (1946), see also Joanes (1972) and Cox's discussion of Armitage (1959), which follows from the asymptotic sufficiency of maximum likelihood estimators. If the sample size is not too small, satisfactory sequential tests might be constructed by using maximum likelihood estimators, assuming the asymptotic distribution.

Suppose that observations $x_1, \ldots, x_n$ are independent and identically distributed random variables with a density $f(x|\theta, \phi)$, and suppose that we wish to test $H_0: \theta = \theta_0$ against $H_1: \theta = \theta_1$, where $\phi$ is a nuisance parameter. If $\phi$ is known, an SPRT is based on

$$L_n(x_1, \ldots, x_n | \theta_1, \phi) - L_n(x_1, \ldots, x_n | \theta_0, \phi), \tag{4.21}$$

where $L_n$ denotes the log-likelihood. (Below we shall write $L_n(\theta, \phi)$ for the log-likelihood of $x_1, \ldots, x_n$.) If (4.21) is expanded as far as quadratic terms about $(\theta, \phi)$, we have

$$(\theta_1 - \theta_0) \frac{\partial L_n(\theta, \phi)}{\partial \theta} + \tfrac{1}{2}(\theta_1 - \theta_0)(\theta_1 + \theta_0 - 2\theta) \frac{\partial^2 L_n(\theta, \phi)}{\partial \theta^2}. \tag{4.22}$$

If $\phi$ is unknown, we can consider basing a test on

$$L_n(x_1, \ldots, x_n | \theta_1, \hat{\phi}) - L_n(x_1, \ldots, x_n | \theta_0, \hat{\phi}), \tag{4.23}$$

where $\hat{\phi}$ is the maximum likelihood estimate. If (4.23) is expanded about the true point $(\theta, \phi)$, we have

$$(\theta_1 - \theta_0) \frac{\partial L_n(\theta, \phi)}{\partial \theta} + \tfrac{1}{2}(\theta_1 - \theta_0)(\theta_1 + \theta_0 - 2\theta) \frac{\partial^2 L_n(\theta, \phi)}{\partial \theta^2}$$
$$+ (\theta_1 - \theta_0)(\hat{\phi} - \phi) \frac{\partial^2 L_n(\theta, \phi)}{\partial \theta . \partial \phi}. \tag{4.24}$$

Thus if and only if $\quad \dfrac{1}{n} \dfrac{\partial^2 L_n(\theta, \phi)}{\partial \theta . \partial \phi} \to 0$

in probability, a test based on (4.23) will be asymptotically identical to a test based on (4.21). This condition means that the maximum likelihood estimators $\hat{\theta}$ and $\hat{\phi}$ must be asymptotically independent.

The maximum likelihood estimates $\hat{\theta}$ and $\hat{\phi}$ satisfy approximately

$$I_{\theta\theta}(\hat{\theta}-\theta)+I_{\theta\phi}(\hat{\phi}-\phi) = \frac{1}{n}\frac{\partial L_n(\theta,\phi)}{\partial\theta}$$

and

$$I_{\theta\phi}(\hat{\theta}-\theta)+I_{\phi\phi}(\hat{\phi}-\phi) = \frac{1}{n}\frac{\partial L_n(\theta,\phi)}{\partial\phi},$$

asymptotically, where

$$I_{\theta\theta} = -E\left\{\frac{\partial^2 \log f(x|\theta,\phi)}{\partial\theta^2}\right\},$$

etc. Further, we can write

$$\frac{\partial^2 L_n(\theta,\phi)}{\partial\theta^2} \sim -nI_{\theta\theta},$$

etc., and, using these equations, it can be shown that (4.24) is asymptotically equivalent to

$$nI_{\theta\theta}(\theta_1-\theta_0)\,\{\hat{\theta}-\tfrac{1}{2}(\theta_1+\theta_0)\}.$$

This indicates that a test can be based on

$$T_n = n\{\hat{\theta}-\tfrac{1}{2}(\theta_1+\theta_0)\}.$$

Now

$$E(T_n) = n\{\theta-\tfrac{1}{2}(\theta_1+\theta_0)\},$$

and

$$V(T_n) = n\Big/\left\{I_{\theta\theta}-\frac{I_{\theta\phi}^2}{I_{\phi\phi}}\right\} = n\tau^2(\theta),$$

say. The process $T_n$ is a random walk with independent increments of mean $\{\theta-\frac{1}{2}(\theta_1+\theta_0)\}$ and variance $\tau^2(\theta)$. Hence under very general conditions, the asymptotic test based on $T_n$ is based on normal distribution theory, and the stopping limits, OC-curve and ASN-function follow from Theorems 2.1, 2.2 and 2.3 respectively. Specifically, $T_n$ is computed until it falls outside the limits

$$\left\{\tau^2(\theta).\frac{\log\left(\dfrac{\beta}{1-\alpha}\right)}{(\theta_1-\theta_0)},\qquad \tau^2(\theta).\frac{\log\left(\dfrac{1-\beta}{\alpha}\right)}{(\theta_1-\theta_0)}\right\},$$

where $\tau^2(\theta)$ must, of course, be estimated from the data.

*Ex. 4.7.* Suppose $x_1, \ldots, x_n$ are independently and normally distributed random variables, with mean $\mu$ and variance $\sigma^2$, both unknown. Suppose further, that we wish to test

$$H_0: \delta = \frac{\mu}{\sigma} = 0 \quad \text{against} \quad H_1: \delta = \delta_1.$$

The maximum likelihood estimate of $\delta$ is $\frac{\bar{x}}{s}$ (except for a factor $n/(n-1)$), in standard notation. A test is carried out by calculating

$$n\left(\frac{\bar{x}}{s} - \frac{\delta_1}{2}\right)$$

with stopping limits

$$\frac{\left(1+\frac{\bar{x}^2}{2\sigma^2}\right)}{\delta_1} \cdot \left\{\log\left(\frac{\beta}{1-\alpha}\right), \log\left(\frac{1-\beta}{\alpha}\right)\right\}.$$

The OC-curve of this test is

$$L(\delta) = \left\{\left(\frac{1-\beta}{\alpha}\right)^h - 1\right\} \Big/ \left\{\left(\frac{1-\beta}{\alpha}\right)^h - \left(\frac{\beta}{1-\alpha}\right)^h\right\}$$

where

$$h = 1 - \frac{2\delta}{\delta_1}.$$

The ASN-function is

$$\left(1+\frac{\delta^2}{2}\right)\left[L(\theta)\log\left(\frac{\beta}{(1-\alpha)}\right) - \{1-L(\theta)\}\log\left(\frac{1-\beta}{\alpha}\right)\right] \Big/ \left\{\delta_1\left(\delta - \frac{\delta_1}{2}\right)\right\}.$$

The OC-curve is identical to that for a test of

$$H_0: \mu = 0 \quad \text{against} \quad H_1: \mu = \delta_1 \sigma$$

if $\sigma^2$ were known, but the ASN-function is $(1+\delta^2/2)$ times the ASN-function of the test for $\sigma^2$ known.

Cox states that this test is asymptotically equivalent and numerically close to the sequential $t$-test given by Rushton (1950).

For a more rigorous justification of Cox's approach as applied to a wide class of sequential tests, see Breslow (1969), who also gives detailed applications to the comparison of two binomial populations, and to the comparison of two exponential survival curves (see P4.7).

An interesting application of this asymptotic test is given by Cox and Roseberry (1966). The observations $y_i$ are the differences between the responses of two alternative medical treatments, taken on paired subjects, but there is a concomitant observation $x$ on each pair. The distribution of $(y, x)$ is assumed to be bivariate normal and it is required to test two hypotheses about $E(y)$. Since the variances and correlation coefficient of the bivariate normal distribution are unknown, we have composite hypotheses, and Cox's asymptotic theory applies. An interesting point here is that some simulation trials of the final test showed that the actual error rates were greatly in excess of those set. The excess error rate was apparently due to premature decisions, and this difficulty was overcome by only recognizing decisions after a given minimum sample size.

Joanes (1972) points out an alternative and asymptotically equivalent version of the above method. The essential point is that he uses maximum likelihood estimates $\hat{\phi}_0$ and $\hat{\phi}_1$ given $\theta_0$ and $\theta_1$ respectively, in (4·23), instead of the estimate $\hat{\phi}$. This leads to a slightly different form of the test. Cox's presentation, however, is simpler.

**Problems 4**

1. Cox suggests that an approximate sequential test for the correlation coefficient be obtained as follows.

If the sample correlation coefficient is denoted $r_n$, then $z_n$

$$z_n = \tfrac{1}{2}\log\frac{1+r_n}{1-r_n}$$

is approximately normally distributed with expectation and variance

$$E(z_n) = \tfrac{1}{2}\log\frac{1+\rho}{1-\rho} \qquad V(z_n) = 1/(n-3)$$

where $\rho$ is the population correlation coefficient. Assume this normal approximation to be exact, and apply Cox's theorem to obtain a sequential test.

2. Use a method similar to that used in P4.1 to obtain an approximate test for the variance of a normally distributed random variable (with unknown mean) by using the logarithmic transformation.

3. Check the mathematics of Ex. 4.1, and determine the asymptotes of the boundaries.

4. Various generalizations of Cox's theorem would appear to be possible. For example, $y_1$ of (4.17) may be a vector of sufficient statistics. Thus for observations $x_i$ distributed independently and normally, we can consider the sequence of statistics

$$\bar{x}_n = \frac{1}{n}\sum_{i=1}^{n} x_i$$

and

$$s_n^2 = \frac{1}{n-1}\sum_{i=1}^{n} (x_i-\bar{x})^2,$$

for observations $x_1, \ldots, x_n$. The likelihood can be factorized into

$$f_1(\bar{x}_n, s_n^2)\, f_2(\bar{x}_1, s_1^2, \ldots, \bar{x}_{n-1}, s_{n-1}^2 | \bar{x}_n, s_n^2),$$

similar to (4.17). Produce a generalization of Cox's theorem to cope with situations of this kind.

5. Suppose observations $x_i$ are independently and normally distributed with known variance unity, and we wish to test $H_0: \mu = 0$ against $H_1: \mu = \pm\mu_1$. Consider the sequence of statistics $n\bar{x}_n^2$ where $\bar{x}_n$ is the sample mean of $n$ observations. Under $H_0$, $n\bar{x}_n^2$ is distributed $\chi_1^2$, whereas under $H_1$ it has a non-central $\chi^2$ distribution. Examine whether Cox's theorem can be applied, and a sequential test obtained along these lines.

6. Various methods have been suggested in this book of formulating a two-sided test of a normal mean, $\sigma^2$ known. (See P4.5 above, § 3.3, § 3.4, Ex. 4.1.) No detailed comparison of these plans is known to the author. Design a set of empirical sampling trials to compare these alternative plans. Discuss in detail how the trials would be carried out, what results you would obtain, etc., and how many trials you would need.

Notice that the plans will have different OC-curves, and so are not strictly comparable. On what basis will you arrange for your tests to be 'equivalent'?

7. Armitage (1959) discussed the application of sequential methods to the comparison of survival curves. Assume that two treatments

for a disease are to be compared, and the measure of success of a treatment is length of survival of the patient after treatment. The probability distribution of survival times can be assumed to be exponential, with parameters $\lambda_A$, $\lambda_B$, for the two treatments. Suppose it is required to test $H_0:\lambda_A/\lambda_B = 1$ against $H_1:\lambda_A/\lambda_B = \delta > 1$, with probabilities of error $(\alpha, \beta)$. The difficulty in forming a sequential test is that at any time $t$ there may be many patients still surviving. Obtain an approximate sequential test based on Cox's asymptotic maximum likelihood method. (An exact sequential test for a single exponential is given by Epstein and Sobel (1955).)

8. The signed-rank (Wilcoxon) test is a fixed sample size test operating as follows. Rank all observations in absolute magnitude,

$$|x_{(1)}| < |x_{(2)}| < \dots < |x_{(n)}|$$

(assume the possibility of ties can be ignored), and form the sum

$$S = \sum r\, d_r$$

where

$$d_r = \begin{cases} +1 & \text{if} \quad x_{(r)} > 0 \\ -1 & \text{if} \quad x_{(r)} < 0 \end{cases}.$$

Assume that the observations $x_i$ are from a symmetrical distribution with unknown mean $\theta$, then for a test of $H_0:\theta = 0$ against alternatives $H_1:\theta > 0$ a test is based on the randomization distribution of $S$ under $H_0$, which is approximately normal. Examine whether an approximate sequential test can be derived using a signed-rank procedure.

9. In the problem discussed in P2.11 put $C = \theta/\phi$ and show that the hypotheses to be tested are $H_0$: $C = C_0$ against $H_1$: $C = C_1$. By considering the sequence

$$Z_r = \sum y_r / \sum x_r$$

and applying Cox's theorem, obtain a sequential test of the hypotheses. (Schafer and Takenaga, 1972).

# CHAPTER 5

# Some Applications of Cox's Theorem

## 5.1. The Sequential $t$-test

In Chapter 4 we derived two alternative forms of the sequential $t$-test, Wald's, Ex. 4.2, and Barnard's, Ex. 4.4, which was obtained by Rushton (1950) based on theory by Barnard (1952).

Now it can be shown that

$$\int_0^\infty \frac{1}{\sigma^n} \exp\left\{-\frac{1}{2\sigma^2} \sum_{i=1}^n (x_i - \delta\sigma)^2\right\} d\sigma$$
$$= \frac{(n-2)! \exp\left\{-\frac{\delta^2}{2}(n - a_n^2)\right\}}{\{\sum x_i^2\}^{\frac{1}{2}(n-1)}} \quad Hh_{n-2}(-\delta a_n) \tag{5.1}$$

where $a_n$ is defined by equation (4.19), and, by using this, the likelihood ratio (4.10) can be expressed in a form compatible with (4.20). If Ex. 4.2 is modified to form a one-sided test between $H_0: \theta = 0$ and $H_1: \theta = \delta_1 \sigma$, then by using (5.1) the likelihood ratio of Wald's test can be obtained in the form

$$\exp\left\{-\frac{\delta_1^2}{2}(n - a_n^2)\right\} . Hh_{n-2}(-\delta_1 a_n)/Hh_{n-2}(0). \tag{5.2}$$

This likelihood ratio differs from the likelihood ratio (4.20) for Barnard's test, in the suffix of the $Hh$-function. However, Cox pointed out that if Wald had used the weight function $1/\sigma$, $0 < \sigma < \infty$, instead of (4.8), the test obtained by Wald would be identical with Barnard's sequential $t$-test. There are no detailed studies of the properties of either Wald's or Barnard's sequential $t$-test for the one-sided case. (Two-sided tests are discussed in § 5.2.)

An outstanding feature of the sequential $t$-tests is our state of ignorance concerning their properties. This derives partly from the point noted earlier, that for the extensions to the SPRT given in

Chapter 4, Wald's method of obtaining approximations to the OC- and ASN-curves usually breaks down. In the case of sequential $t$-tests, the complexity of the mathematics is such that theoretical analysis is unlikely to be fruitful on these points.

Wald's SPRT for $\sigma^2$ known has the lowest possible ASN at $H_0$ and $H_1$ (Wald–Wolfowitz theorem), so that at least at these two points the ASN of the corresponding $\sigma^2$ known SPRT is a lower bound to the ASN of the sequential $t$-test. However, Bartholomew (1956) has indicated that a lower bound obtained in this way is not necessarily a close one. In fact Cox's work on asymptotic maximum likelihood tests (§ 4.7, and Ex. 4.7), indicates that in some asymptotic sense, the ASN of the sequential $t$-test is $(1+\frac{1}{2}\delta^2)$ times the ASN of the $\sigma^2$ known SPRT.

Other known properties of the tests are that they terminate with probability one (Cox, 1952a; David and Kruskal, 1956), and that the probability of accepting $H_0$ is a decreasing function of $\delta$ (or of $\delta^2$). Also, Wald's test possesses a certain minimax property related to the probabilities of error. See Ghosh (1960, 1962) for a discussion of these points.

## 5.2. Two-sided Tests

The two main ways of obtaining a two-sided sequential $t$-test have already been foreshadowed. They are:

(i) superimpose two one-sided tests, as illustrated in § 3.2 – § 3.4;

(ii) reformulate the hypotheses so as to work on $t^2$ rather than $t$. This automatically produces a two-sided test, and Wald's (1947) derivation was of this form (see Ex. 4.2).

Both Barnard's and Wald's tests could be used in either way to produce a two-sided test. The ratio of $Hh$-functions in (4.20), and (4.10, using 5.1), can be expressed as confluent hypergeometric functions, and from this it follows that Barnard's two-sided form (ii) test rejects more frequently than Wald's form (ii) test, see Hoel (1954). Thus although the tests are designed to satisfy the same restrictions on the probabilities of error, the actual probabilities of error cannot be equal.

Arnold (1951) carried out a small empirical sampling experiment to provide a rough comparison of Wald's and Barnard's test for case (ii), and the results are summarized briefly in Table 5.1. The hypotheses tested were $H_0:\theta=0$ and $H_1:|\theta|=\sigma$, with set probabilities of error $\alpha=\beta=0{\cdot}05$. There appears to be little to choose between the two procedures except that Barnard's test has a slightly lower ASN on the alternative hypothesis, and this would be expected from the above discussion.

TABLE 5.1. Empirical comparison of Wald's and Barnard's sequential $t$-tests.†

| Sample Size | Mean Zero | | | |
|---|---|---|---|---|
| | Wald | | Barnard | |
| | Accept | Reject | Accept | Reject |
| 5–10 | 329 | 14 | 321 | 18 |
| 11–15 | 112 | 3 | 109 | 3 |
| 16–25 | 35 | 2 | 42 | 1 |
| > 25 | 5 | 0 | 6 | 0 |
| Total | 481 | 19 | 478 | 22 |
| Average sample size | 9·87 | 10·47 | 10·06 | 9·32 |
| Combined average | 9·89 | | 10·03 | |

| Sample Size | Mean One | | | |
|---|---|---|---|---|
| | Wald | | Barnard | |
| | Accept | Reject | Accept | Reject |
| 5–10 | 14 | 240 | 13 | 281 |
| 11–15 | 1 | 139 | 2 | 114 |
| 16–25 | 3 | 87 | 1 | 77 |
| > 25 | 0 | 16 | 1 | 11 |
| Total | 18 | 482 | 17 | 483 |
| Average sample size | 9·61 | 12·28 | 9·76 | 11·25 |
| Combined average | 12·18 | | 11·20 | |

† These results were extracted from National Bureau of Standards (1951).

Further empirical sampling trials of Barnard's two-sided (case (ii)) sequential $t$-test have been carried out by Schneiderman and Armitage

(1962b), and Siskind (1964), and extracts from their results are given in Table 5.2 and Table 5.3 respectively. The two sets of results agree

TABLE 5.2. Empirical sampling trials of Barnard's sequential $t$-test. Average sample number and operating characteristics†

| $\delta$ | No. of trials | Sample size: Average | Sample size: Standard deviation | Proportion with $H_0$ accepted |
|---|---|---|---|---|
| $\alpha = \beta = 0{\cdot}05$, $\delta_1 = 0{\cdot}5$ | | | | |
| 0·0 | 500 | 35·58 | 13·4 | 0·964 |
| 0·125 | 200‡ | 42·01 | 24·5 | 0·878 |
| 0·250 | 600‡ | 50·87 | 34·1 | 0·551 |
| 0·375 | 200‡ | 46·22 | 29·1 | 0·192 |
| 0·500 | 600 | 32·60 | 19·5 | 0·025 |
| $\alpha = \beta = 0{\cdot}05$, $\delta_1 = 1{\cdot}0$ | | | | |
| 0·0 | 500 | 10·00 | 4·4 | 0·962 |
| 0·25 | 500‡ | 12·31 | 7·0 | 0·860 |
| 0·50 | 500‡ | 14·62 | 9·1 | 0·536 |
| 0·75 | 500‡ | 14·04 | 7·3 | 0·182 |
| 1·00 | 500 | 11·27 | 5·2 | 0·012 |

† Extracted from Schneiderman and Armitage (1962b).
‡ Some of these trials were terminated at $n = 151$ for $\delta = 0{\cdot}5$ trials and $n = 41$ for $\delta = 1{\cdot}0$ trials.

TABLE 5.3. Empirical sampling trials of Barnard's two-sided (form (ii)) sequential $t$-test
$H_0 : \delta_0 = 0$, $H_1 : \delta_1 = 1$, $\alpha = \beta = 0{\cdot}05$†

| $\delta$ | ASN | St. dev. ASN | Proportion with $H_0$ accepted | No. of trials |
|---|---|---|---|---|
| 0·0 | 9·90 | 4·21 | 0·960 | 500 |
| 0·2 | 11·21 | 6·07 | 0·932 | 600 |
| 0·4 | 14·34‡ | 9·33 | 0·746 | 700 |
| 0·6 | 15·25‡ | 9·72 | 0·363 | 700 |
| 0·8 | 13·05‡ | 7·38 | 0·113 | 600 |
| 1·0 | 10·97 | 4·66 | 0·050 | 500 |

† Extracted from Siskind (1964).
‡ Some of these trials were terminated at $n = 61$.

very well except for $\delta = 1 \cdot 0$ on the test with $\delta_1 = 1 \cdot 0$, and here the results of Tables 5.1, 5.2 and 5.3 should all agree. In fact, while the ASN's agree well, there are large differences in the proportion of trials accepting $H_0$, and the reason for this is not clear.

Siskind (1964) also tabulates the distribution of the decisive sample number, and we find that when $\delta = 0 \cdot 6$, when the observed ASN was 15·25, four of the seven hundred trials were terminated at $n = 61$ without a decision having been made. See also Table 5.1 for further information on the sample size distribution, which clearly has a very long tail, so that the observed standard deviations need to be interpreted with care.

## 5.3. Closed Forms of the Tests

All forms of the sequential $t$-test described so far are based on the SPRT, and so yield open boundaries. The infinite range of sample sizes possible on open plans frequently gives rise to administrative difficulties, and in many practical situations experimenters prefer to use closed tests, see Armitage (1960).

Schneiderman and Armitage (1962b) have proposed two forms of closed sequential $t$-test, based on their earlier work (1962a) on tests for a normal mean when the variance is known. (This work will be described in Chapter 6.) The proposals are to remove the boundary of the SPRT at which $H_0$ is accepted, and replace it by either

(a) 'Restricted procedures.' Acceptance of $H_0$ as soon as a given sample size is reached.

or (b) 'Wedge plans.' A boundary for acceptance of $H_0$ which is a quadratic (increasing) function of $n$.

These boundaries are illustrated in Fig. 5.1, which is similar to that of Schneiderman and Armitage (1962b).

Some empirical sampling trials were run to compare the two closed plans with Barnard's open sequential $t$-test. Unfortunately, comparisons were vitiated because the open plan had greater power than either of the closed plans, but the closed plans are not very successful. Restricted plans have a very poor ASN for values of $\delta$ near $H_0$, and this is to be expected from the shape of the boundaries. If the difference in power is ignored, wedge plans have a similar ASN to open plans, and may have a slightly smaller ASN than the open plans for some range of $\delta$ intermediate to $H_0$ and $H_1$. The observed variances of the sample size distributions of the closed plans were mostly smaller than for the open plan, and sometimes very much smaller.

Suich and Iglewicz (1970) have also developed a closed sequential $t$-test based on the work of Anderson (1960), which is described in Section 6.6. Basically, Anderson uses boundaries for the log-likelihood which are straight lines against $n$, and converge as $n$ increases. Suich and Iglewicz simply use Anderson's boundaries, which are derived for the normal distribution, for the log-likelihood obtained in the sequential $t$-test. Only the case $\alpha=\beta$ is considered, and various simple properties of the resulting tests are obtained by empirical sampling and compared with properties of corresponding versions of Barnard's sequential $t$-test, and Schneiderman and Armitage's test described above. The work is extended to the case $\alpha\neq\beta$ by Alexander and Suich (1973). The general conclusions are that the new boundaries may be preferred on the basis of ASN to both Barnard's and Scheiderman and Armitage's test. However there is a practical difficulty that only a few sets of boundaries have been worked out for any of the closed procedures. Furthermore there are differences in power which make comparisons rather difficult (Barnard's test is more powerful than the others).

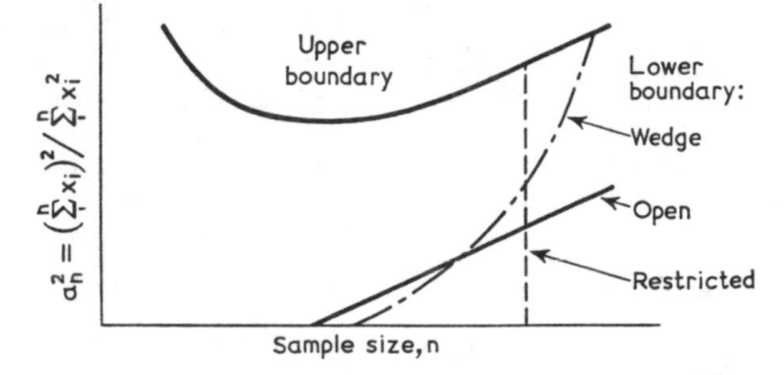


FIG. 5.1. Boundaries for three two-sided (form (ii)) sequential $t$-tests.

### 5.4. Use of the Tests

Armitage (1947) evaluated the integrals (4.10), and produced a short table containing four sets of boundaries for Wald's two-sided (form (ii)) test. The National Bureau of Standards (1951) produced tables which give the boundaries for Barnard's two-sided (form (ii)) test, for a large range of values of the parameters. These tables can also be used to obtain the boundaries for Wald's two-sided (form (ii)) test.

For the one-sided test, Rushton calculated boundaries for Barnard's one-sided sequential $t$-test, and the tables have been incorporated in Davies (1954). Rushton (1950, 1952) gave approximations suitable for carrying out both the one-sided and the form (ii) two-sided versions of Barnard's sequential $t$-test.

Davies (1954) gave a numerical illustration of the use of Barnard's one-sided sequential $t$-test, and Hajnal, Sharp and Popert (1959) give an illustration of the use of Barnard's form (ii) two-sided test in a sequential medical trial. (Further details of Hajnal *et al.* experiment and some discussion on the use of the sequential $t$-test in such experiments are given in Hajnal (1960). Other sequential medical trials using a sequential $t$-test include Sainsbury and Lucas (1959), who used Barnard's form (ii) two-sided test, and Kilpatrick and Oldham (1954), who used Barnard's one-sided sequential $t$-test. The latter trial terminated with acceptance of the null hypothesis so quickly that the new treatment was thought to be inferior to the standard, but unfortunately, because of the design used, this supposition could neither be tested nor estimated. All of these applications refer to tests of a single population mean (or paired differences), and not to tests of differences between two or more samples.

A two-sided two-sample sequential $t$-test has been discussed by Hajnal (1961). Hajnal's procedure assumed that at each stage in the sampling, prescribed (constant) numbers $n_1$ and $n_2$ of observations are taken from the two populations. Hajnal's theory is a direct application of Cox's theorem and his test can be carried out using the National Bureau of Standards (1951) tables.

## 5.5. A Sequential Test of Randomness

Cox (1955) gave a detailed discussion of methods of statistical analysis suitable for series of events, such as slubs along the length of a yarn, successive stops of a machine, etc. One of the problems Cox mentioned was to test whether such events are distributed randomly in time or space. Under the hypothesis of randomness, the probability of an event in any interval $\delta t$ (of time or space), is $\mu\delta t$, independent of other events, for some value of $\mu$. This implies that the intervals $t_1, t_2, \ldots, t_n$ between successive events are distributed independently with a probability density function

$$\mu e^{-\mu t}, \qquad t \geqslant 0. \tag{5.3}$$

If the hypothesis of randomness does not hold, there are numerous possibilities, and Cox (1955) discusses eight common types of departure from randomness. Even if the events, such as machine stops, are not distributed randomly, it is still possible that the intervals $t_i$ are statistically independent of each other, for example, the events may tend to occur at intervals near a given length. In some circumstances, therefore, a suitable alternative hypothesis may be that the intervals $t_i$ have a gamma density,

$$\frac{\mu}{\Gamma(\alpha)}(\mu t)^{\alpha-1}\,\mathrm{e}^{-\mu t}, \qquad t \geqslant 0. \tag{5.4}$$

If $\alpha = 1$, the distribution (5.4) reduces to (5.3). The parameter $\alpha$ determines the form of the distribution, while the parameter $\mu$ is merely a scale constant on $t$. The null hypothesis $H_0: \alpha = 1$ that events are randomly distributed can be tested against the alternative hypothesis $H_1: \alpha = \alpha_1$, say, and since $\mu$ is usually unknown, it enters as a nuisance parameter in a rather similar way to that in which $\sigma$ enters into the discussion of the sequential $t$-test.

Cox notes that a test of (5.3) against (5.4) will be insensitive to many common types of departure from randomness, and in particular to trend. The reason for this is that both hypotheses assume that the intervals $t_i$ are statistically independent, and a test of (5.3) against (5.4) will be based on the sample distribution of $t_i$'s, ignoring their order of occurrence.

Bartholomew (1956) obtained a sequential test of (5.3) against (5.4), for $\mu$ unknown, by proceeding in the manner indicated by Cox's theorem. The first requirement is to transform the $t_i$ so as to obtain a sequence of statistics which are independent of $\mu$, and the simplest set of transformations is probably

$$y_2 = \frac{t_1 t_2}{(t_1+t_2)^2}, \quad y_3 = \frac{t_1 t_2 t_3}{(t_1+t_2+t_3)^3}, \quad \ldots, \quad y_n = \frac{\prod_{i=1}^{n} t_i}{(\sum t_i)^n},$$

$$Y = \sum_{i=1}^{n} t_i. \tag{5.5}$$

Cox's theorem can be applied in this situation. The statistics $(y_n, Y)$ are functionally independent, and jointly sufficient for $(\alpha, \mu)$. The functions $(u_1, u_2, \ldots, u_m)$ of Cox's theorem can be equated to $(y_2, \ldots, y_{n-1})$ and these are also functionally independent. The statistics $(y_2, \ldots, y_n)$ are unchanged by the transformation $(t' = at)$, and since $\mu$ is a scale constant this implies $(y_2, \ldots, y_n)$ are independent of $\mu$. The reader should check carefully for himself that the conditions of Cox's theorem are satisfied. The likelihood ratio of $H_1$ to $H_0$ is therefore

$$\frac{f(y_n|\alpha_1)}{f(y_n|\alpha = 1)},$$

where $f(y_n|\alpha)$ is the marginal distribution of $y_n$. Bartholomew showed that this likelihood ratio is

$$R(\alpha_1, n) = \frac{y_n^{\alpha_1 - 1}\,\Gamma(n\alpha_1)}{\Gamma^n(\alpha_1)\,\Gamma(n)}, \tag{5.6}$$

and this ratio is used in connection with the usual limits $(A, B)$ to form an SPRT (see § 2.2). By taking logarithms of (5.6), the SPRT can be expressed in the more convenient form

$$K_1 < \sum \log t_i - n\log\left(\sum t_i\right) < K_2,$$

where the constants $K_1$ and $K_2$ depend on $\alpha_1$, $n$, and the probabilities of error $\alpha, \beta$ of the SPRT.

In order to form this SPRT, a particular value $\alpha_1$ for the alternative hypothesis has to be chosen, parallel to the choice of $\delta_1$ for the sequential $t$-test (see § 4.5). The choice of $\alpha_1$ will often be rather difficult, especially since the alternative (5.4) may be artificial, and used simply to construct the test. Further, owing to the fact that little is known of the properties of Bartholomew's test, the consequences of a particular choice for $\alpha_1$ are not easy to examine.

Since the $y_i$ (5.5) are not independent, no optimum property can be stated for the test. However, Bartholomew has an interesting method of obtaining an OC-curve, and the method may be applicable to other sequential tests involving dependent variables.

The method used in § 2.3 for obtaining the OC-curve (Theorem 2.2), works out as follows for the present test. Denote the probability density of $(y_2, \ldots, y_n)$ given $\alpha$ by $p(y_2, \ldots, y_n|\alpha)$, then parallel to the proof of Theorem 2.2 we must find an $h$ depending on $\alpha$ but not on $n$, satisfying

$$\int \cdots \int \left\{\frac{p(y_2, \ldots, y_n|\alpha = \alpha_1)}{p(y_2, \ldots, y_n|\alpha = 1)}\right\}^h p(y_2, \ldots, y_n|\alpha)\,dy_2 \ldots dy_n = 1. \tag{5.7}$$

If the solution depends on $n$, the proof of the theorem breaks down, for the limits of the modified SPRT (2.10) would not then satisfy (2.11) and (2.12). When $h = 1$, then $\alpha = 1$ satisfies (5.7) for all $n$, and similarly, when $h = -1$ then $\alpha = \alpha_1$ for all $n$. For $|h| \neq 1$, the $\alpha$ satisfying (5.7) will in general depend on $n$, but under certain conditions the function $h(\alpha, n)$ can be shown to be independent of $n$ to a first approximation.

Write the left-hand side of (5.7) as a function

$$G_n(x, z, \alpha) = \int \cdots \int \left\{\frac{p(y_2, \ldots, y_n|\alpha = x)}{p(y_2, \ldots, y_n|\alpha = z)}\right\}^h \times p(y_2, \ldots, y_n|\alpha)\, \mathrm{d}y_2, \ldots, \mathrm{d}y_n$$

so that (5.7) is $G_n(\alpha_1, 1, \alpha)$. Under certain conditions the function $G_n$ can be expanded into a Taylor's series about the point $G_n(\alpha, \alpha, \alpha)$. The integrals involved in the expansion up to the second-order terms are

$$E\left\{\frac{\mathrm{d} \log p(y_2, \ldots, y_n|\alpha)}{\mathrm{d}\alpha}\right\}, \tag{5.8}$$

$$E\left\{\frac{\mathrm{d}^2 \log p(y_2, \ldots, y_n|\alpha)}{\mathrm{d}\alpha^2}\right\} \tag{5.9}$$

and

$$E\left\{\frac{\mathrm{d} \log p(y_2, \ldots, y_n|\alpha)}{\mathrm{d}\alpha}\right\}^2. \tag{5.10}$$

Further, under general conditions the expectation (5.8) is zero, and (5.9) is equal to minus (5.10) (see Cramer, 1946, Chapter 32) and (5.7) reduces to

$$h\left[\left(\frac{h-1}{2}\right)\{(\alpha_1 - \alpha) - (1-\alpha)\}^2 + (1-\alpha)^2 - (\alpha_1 - \alpha)(1-\alpha)\right] \times E\left(\frac{\mathrm{d} \log p(y_2, \ldots, y_n|\alpha)}{\mathrm{d}\alpha}\right)^2 + \text{terms of third order} = 0.$$

If terms of third order can be neglected, then

$$h \simeq 1 - 2\left(\frac{\alpha - 1}{\alpha_1 - 1}\right), \tag{5.11}$$

and this equation is exact at $\alpha = 1$ and $\alpha = \alpha_1$.

This argument is of quite general application and could be used for other sequential tests where Cox's theorem is involved. The critical condition, however, is that the third-order terms in the expansion can be neglected, and even where this is true it is difficult to demonstrate. For the test of randomness of intervals, Bartholomew is able to show that the value of $\alpha$ for which $h(\alpha, n) = 0$ is also nearly independent of $n$. (An exact equation for $\alpha$ as a function of $n$, satisfying $h(\alpha, n) = 0$, can be obtained.)

Thus at $h = 1$ or $h = -1$, $\alpha$ is independent of $n$, and at $h = 0$, $\alpha$ is nearly independent of $n$. Bartholomew argues that this is reasonable evidence to assume the solution of (5.7) to be nearly independent of $n$. It follows that Theorem 2.2 provides a method of obtaining an approximate OC-curve. The approximations involved are firstly ignoring the overshooting of the boundaries, and secondly, neglecting terms of third and higher order in the expansion of (5.7). Unfortunately, Wald's method of obtaining upper and lower bounds for the OC-curve fails for a sequence of dependent variables, and there is at present no way of setting bounds on the error introduced by these approximations. However, the OC-curve obtained by the above procedure could be compared with empirical sampling trials in a few cases, as a check.

When the above method is applied to Bartholomew's sequential test of randomness, the approximate OC-curve for a test of $H_0: \alpha = 1$ against $H_1: \alpha = \alpha_1$, where $\mu$ is unknown, is identical with the SPRT of these hypotheses where $\mu$ is known, for the same prescribed probabilities of error.

Bartholomew obtains an approximate ASN for his test by using a conjecture of Bhate (1955). Let us denote the probability of accepting the null hypothesis when $\alpha$ is true by $L(\alpha)$, then we have approximately

$$E\{\log K(\alpha, n)\} = L(\alpha) \log B + \{1 - L(\alpha)\} \log A, \qquad (5.12)$$

where $A$ and $B$ are the limits for the SPRT (see § 2.2), and $K(\alpha,n)$ is the likelihood ratio, given by (5.6) for the present case. Bhate suggests that $n$ in $K(\alpha,n)$ be replaced by $E(n|\alpha)$, and then (5.12) is solved for $E(n|\alpha)$. Again, there is no method of checking the approximation involved in this procedure, and some empirical sampling trials are desirable.

Siskind (1964) studied both Bartholomew's idea for approximating to the OC-curve of an SPRT on dependent observations, and also Bhate's conjecture concerning the ASN-function, with respect to

Barnard's two-sided form (ii) sequential $t$-test. The empirical sampling trials summarized in Table 5.3 were part of a programme to check the accuracy of these approximation procedures, and the general conclusions are that the approximations are satisfactory only for certain ranges of the parameter under test.

## 5.6. Sequential Analysis of Variance

Suppose we have $g$ bobbins of yarn, and we wish to test the null hypothesis that they all have a common population mean breaking strength. If the variance within bobbins can be assumed to be homogeneous, then the null hypothesis can be tested by a one-way fixed effects analysis of variance. In a standard fixed sample size approach we should try to obtain an estimate of the error variance in order to estimate the number of observations necessary to achieve a desired power. If no estimate of the error variance is available, a fixed sample size design runs into severe risk of either an inconclusive experiment or, alternatively, waste of effort in that a much smaller number of observations than those actually taken may have been sufficient. Further, if observations are costly in money, time or material, then a sequential design for testing the null hypothesis may be appropriate, provided it is reasonably simple to carry out.

One way of forming a sequential design is as follows. The experiment is carried out in stages, and at each stage a fixed number $n_i$, for $i = 1, 2, \ldots, g$, of observations are taken from each group (bobbin). Denote the $j$th observation on the $i$th group at the $r$th stage by $x_{ijr}$, then we calculate

$$CS_{b,r} = \sum_{i=1}^{g} n_i(\bar{x}_{i.r} - \bar{x}_{..r})^2$$

and

$$CS_{w,r} = \sum_{i=1}^{g} \sum_{j=1}^{n_i} (x_{ijr} - \bar{x}_{i.r})^2,$$

where

$$\bar{x}_{i.r} = \frac{1}{n_i} \sum_{j=1}^{n_i} x_{ijr} \quad \text{and} \quad \bar{x}_{..r} = \frac{1}{\sum n_i} \sum_{i=1}^{g} \sum_{j=1}^{n_i} x_{ij}.$$

Thus we have a sequence of corrected sums of squares

$$(CS_{b,1}, CS_{w,1}), (CS_{b,2}, CS_{w,2}), \ldots, (CS_{b,N}, CS_{w,N}),$$

and if the $x_{ijr}$ are independently and normally distributed, all the terms of this sequence are statistically independent.

Suppose the $F$-ratios are calculated at each stage,

$$F_r = \frac{CS_{b,r}/(g-1)}{CS_{w,r}\Big/\sum_{i=1}^{g}(n_i-1)} \tag{5.13}$$

for $r = 1, 2, \ldots, N$. For fixed effects analysis of variance, the terms $F_r$ are independently and identically distributed with a non-central $F$-distribution, and non-centrality parameter

$$\lambda^2 = \sum_{i=1}^{g} n_i(\mu_i - \bar{\mu})^2/(g-1)\,\sigma^2 \tag{5.14}$$

where $\mu_i$ is the true mean for the $i$th group, $\bar{\mu}$ is the average of the $\mu_i$, and $\sigma^2$ the error variance. (See Patnaik (1949), for a statement of the non-central $F$-density, but note that in Patnaik's formula, his $\delta = \lambda^2$ of (5.14).)

Thus to test the hypotheses

$$H_0 : \lambda^2 = 0 \quad \text{against} \quad H_1 : \lambda^2 = \lambda_1^2,$$

the theory of Chapter 2 applies directly. However, this procedure is complicated to follow through.

Another procedure (there are many more), is to recompute the between and within groups sums of squares at each stage,

$$CS'_{b,r} = r\sum_{i=1}^{g} n_i(\bar{x}_{i..} - \bar{x}_{...})^2,$$

and

$$CS'_{w,r} = \sum_{i=1}^{g}\sum_{j=1}^{n_i}\sum_{s=1}^{r}(x_{ijs} - \bar{x}_{i..})^2,$$

and compute the sequence of $F$-ratios

$$F'_r = \frac{CS'_{b,r}/(g-1)}{CS'_{w,r}\Big/\left\{r\sum_{i=1}^{g} n_i - g\right\}}. \tag{5.15}$$

The simplest case to consider is $n_i = 1$ for all $i$; that is, one more observation is taken from each group at each stage.

The sequence of $F$-ratios (5.15) is not independent, and Cox's theorem must be applied. An SPRT can be formed, depending only on the last $F$-ratio calculated, and for testing

$$H_0 : \lambda^2 = 0 \quad \text{against} \quad H_1 : \lambda^2 = \lambda_1^2,$$

where $\lambda^2$ is the non-centrality parameter (5.14), the likelihood ratio is a ratio of non-central to central $F$-distributions.

For the random effects model analysis of variance, the ratios (5.13) and (5.15) have probability densities proportional to the central $F$-distribution, and the procedures are much simpler. Unfortunately, these procedures do not always terminate with probability one (see P5.6).

Several criticisms can be levelled at sequential analysis of variance, apart from the complicated nature of the resulting test. Firstly, the tests are based on hypotheses about quantities such as (5.14), which do not usually have any practical meaning, but are introduced merely in order to formulate an hypothesis which can be tested precisely. Secondly, some analyses of variance are carried out to 'spot the winner', etc., and in such cases some groups could possibly be dropped from testing at an early stage. The procedures above require all groups to be included until trials are terminated.

For details of some sequential analysis of variance procedures and some other work related to it see Ghosh (1967), Johnson (1953), and Ray (1956).

## 5.7. Sequential $\chi^2$ and $T^2$

Jackson and Bradley (1961a), constructed sequential tests for a multivariate normal population. Let $\underline{\mu} = (\mu_1, \ldots, \mu_p)$ be the vector of means and $\underline{\Sigma}$ the covariance matrix of the multivariate normal distribution, and let the sample estimates of these be $\underline{\bar{x}}_n$, and $\underline{S}_n$ respectively.

If it is required to test the hypothesis

$$H_0 : (\underline{\mu} - \underline{\mu}_0)\, \underline{\Sigma}^{-1} (\underline{\mu} - \underline{\mu}_0)' = \lambda_0^2 \tag{5.16}$$

against the alternative

$$H_1 : (\underline{\mu} - \underline{\mu}_0)\, \underline{\Sigma}^{-1} (\underline{\mu} - \underline{\mu}_0)' = \lambda_1^2 \tag{5.17}$$

then tests may be based on calculating at each sample size $n$,

$$\chi_n^2 = (\underline{\bar{x}}_n - \underline{\mu}_0)\, \underline{\Sigma}^{-1} (\underline{\bar{x}}_n - \underline{\mu}_0)^1 \tag{5.18}$$

if $\underline{\Sigma}$ is known, and

$$T_n^2 = (\underline{\bar{x}}_n - \underline{\mu}_0)\, \underline{S}^{-1} (\underline{\bar{x}}_n - \underline{\mu}_0)^1 \tag{5.19}$$

if $\underline{\Sigma}$ is not known. The sequences of statistics (5.18) and (5.19) are not independent.

By a direct application of Cox's theorem, SPRT's can be formed for testing the hypotheses (5.16) and (5.17), the tests being based on the last value of (5.18) or (5.19) computed. However, Jackson and Bradley state, 'Verification of the conditions of Cox's theorem has not been included by other authors using the theorem. Since we have not found the necessary verifications trivial either for this paper or others already published, we include a sketch of the required demonstrations.'

The reader is referred to the theory described in Jackson and Bradley (1961a), and the application, described in Jackson and Bradley (1961b).

**Problems 5**

1. Examine whether Bartholomew's method of obtaining an approximate OC-curve for dependent random variables is applicable to the one-sided sequential $t$-tests. (See Siskind, 1964.)

2. Suppose that you have an electronic computer available, and can carry out empirical sampling trials on sequential $t$-tests. List the properties of one- and two-sided $t$-tests you would investigate. Plan a series of computer runs for (a) comparing Barnard's and Wald's one-sided test, and (b) comparing alternative forms of two-sided tests. (See P4.6.)

3. Plot the boundaries of alternative one- and two-sided $t$-tests. (For example, take $H_0: \delta = 0$ against $H_1: \delta = \pm 0{\cdot}1$, with $\alpha = \beta = 0{\cdot}05$.) Examine asymptotic shapes of the boundaries. (For some approximations to the boundaries, see Rushton (1950) and (1952) and Armitage (1947).) Discuss what qualitative statements you can make about the alternative forms of test. Why will these statements be of limited value?

4. The sequential $t$-tests all relate to hypotheses about $\delta = \mu/\sigma$. Under what circumstances would it be preferable to test hypotheses about $\mu$? Obtain an approximate one-sided SPRT for testing hypotheses about $\mu$, by using Cox's asymptotic maximum likelihood method described in § 4.7. Compare the boundaries with those of Ex. 4.7. (See Cox (1963).)

5. Obtain an approximate sequential $t$-test as follows. Assume the parameter of interest is $\delta = \mu/\sigma$, in standard notation. Assume further that $\delta$ is approximately normally distributed with the appropriate mean and variance, and construct an approximate test on this basis. (See Cox, 1952a.) When would you expect the approximation to be poor?

6. Suppose you have sequential analysis of variance for the random effects model, with just one-way classification. Initially, $g$ groups are chosen at random. Then at each stage one observation is chosen from each group and the quantity (5.15) calculated. The distribution of the ratio (5.15) depends on the ratio, say $\delta$, of the components of variance between and within groups. Assume Cox's theorem applies, and obtain the SPRT for testing $H_0 : \delta = \delta'$, against $H_1 : \delta = \delta''$, with risks $(\alpha, \beta)$. Put the SPRT in the form

$$R_{1,n} < F'_n < R_{2,n}$$

where $R_{1,n}$ and $R_{2,n}$ depend on $\delta'$ and $\delta''$, and the risks $(\alpha, \beta)$. Show that unless $\delta' = 0$,

$$\lim_{n \to \infty} R_{1n} \neq \lim_{n \to \infty} R_{2,n}$$

where $n$ is the number of observations taken from each group. Hence infer that if $\delta' \neq 0$, the probability of the SPRT terminating is not always unity. (See Johnson, 1953, p. 619.)

7. Study the practical applications of sequential $t$-tests referred to in the text, and then examine critically the formulation of the sequential $t$-test. (See Hajnal (1960).) Would a sequential estimation procedure be more useful? The sequential $t$-test is the question for which an answer can be obtained and is known. What question or questions are likely to occur more often in practice?

8. Consider the problem of extending the methods of Section 5.5 to deal with a sequential test of $H_0$: $\alpha = \alpha_0$ versus $H_1$: $\alpha = \alpha_1$, in the gamma distribution (5.4) with $\mu$ unknown (see Phatarfod, 1971).

CHAPTER 6

# Some Methods Leading to Closed Boundaries

## 6.1. Review of Chapters 1–5

The body of material presented in Chapters 1 to 5 inclusive is very interesting from a theoretical viewpoint. Practical applications, however, have not been as plentiful as might have been hoped, and the search for examples has had to be extensive (although published applications presumably represent a small fraction of the total).

On the other hand some techniques such as double and multiple sampling have been in widespread use for a long period. There is a marked reluctance to apply the SPRT as such, and probably two factors are involved:

(a) Having to check after every observation for a terminal decision is often considered unduly complicated, and in some applications would be wasteful, as in the whitefish sampling problem, Ex. 2.2.

(b) Variability in the sample size can only be tolerated within limits.

The SPRT arises largely because of mathematical convenience, and because it frequently leads to a relatively simple technique. For most practical situations, such as the applications listed in this book, the tests required do not naturally arise as tests of simple hypotheses, and the optimality property stated by the Wald–Wolfowitz theorem is not often relevant. The sample size distribution yielded by the SPRT has a greater spread than most experimenters appear to be willing to tolerate. In this chapter we review the attempts by various authors to generate tests more appropriate to their needs.

## 6.2. Armitage's 'Restricted' Procedures

Armitage (1960) has been concerned with the design of experiments for the comparison of alternative medical treatments. He argues that ethical considerations demand a sequential trial, but that the infinite range of sample sizes possible on an SPRT could not be tolerated (see

Armitage, 1957, pp. 9–10). As an alternative to the SPRT he devised some 'restricted' sequential procedures, which we study here.

Consider a two-sided test of a normal mean $\mu$, when the variance $\sigma^2$ is known, to test hypotheses

$$H_0 : \mu = 0, \qquad H_1 : |\mu/\sigma| > \delta_1.$$

Let the observations be $x_i$, and put

$$y_n = \sum_1^n x_i.$$

Armitage suggested we consider straight-line boundaries, with two lines, U and L, placed symmetrically with respect to the axis $y = 0$, and a definite upper limit $N$ to the sample size. Write the boundaries,

$$\left.\begin{aligned} \mathrm{U}: y_n &= a + bn \\ \mathrm{L}: y_n &= -a - bn \end{aligned}\right\} (a, b > 0). \tag{6.1}$$

The hypothesis $H_1$ is accepted if U or L is reached, and $H_0$ is accepted if the sample size $N$ is reached. (Armitage allowed for some boundaries with $b < 0$, but his approximations are questionable for this case, and in practical applications he used $b > 0$.)

Armitage suggests choosing plans of this type which satisfy the following restrictions:

(i) If $\mu = 0$, the probability of a path ending on either U or L is $\alpha$, so that the probability of a path ending with acceptance of $H_0$ is $(1-2\alpha)$.

(ii) If $\mu = \delta_1 \sigma$ (or $-\delta_1 \sigma$) the probability of reaching U (or L) is $1-\beta$.

Now the 'restricted' sequential plans have three parameters, $(a, b, N)$, so that one more condition could be set. Armitage suggests that along the upper boundary U, the likelihood ratio for $\mu = \delta_1 \sigma$ to $\mu = 0$ be made constant and equal to $(1-\beta)/\alpha$.† This leads to choosing

$$\left.\begin{aligned} a &= \sigma \log\{(1-\beta)/\alpha\}/\delta_1 \\ b &= \delta_1 \sigma/2 \end{aligned}\right\}. \tag{6.2}$$

This last restriction has certain mathematical advantages, which are outlined below.

The plans are now completely specified in principle, and a unique set $(a, b, N)$ corresponds to any given pair of probabilities of error

† The U and L boundaries coincide with those specified in § 3.2 (Ex. 3.1 and Fig. 3.2), but the $\alpha, \beta$ there refer to different probabilities.

$(\alpha,\beta)$. It is important to notice that no optimal quality is claimed for the plan. It is presumably a compromise between the kind of boundaries thought to be desirable, and those for which the mathematical problems posed can be solved.

### 6.3. Approximate Solution by Diffusion Theory

Denote by $P(\delta|a,b,N)$ the probability of reaching the upper boundary when the mean $\mu = \delta\sigma$. The likelihood ratio on the upper boundary having been fixed at $(1-\beta)/\alpha$ for $\mu = \delta\sigma$ to $\mu = 0$, it follows that if we make

$$P(\delta_1|a,b,N) = 1-\beta \tag{6.3}$$

then we must have

$$P(0|\alpha,b,N) = \alpha.$$

Now $a$ and $b$ are given in terms of $\alpha$, $\beta$ and $\delta_1$ by (6.2), so that if $\alpha$, $\beta$, $\delta_1$ are given, we only have to solve (6.3) for $N$.

An approximation to $P(\delta|a,b,N)$ can be obtained by diffusion theory. Let us ignore the lower boundary, and approximate discrete movements in the sample size $n$ by continuous movements in time. The sloping boundary can be represented by a horizontal absorbing boundary at $y = a$, and drift $(\delta\sigma - b)$ per unit time, and growth in variance $\sigma^2$ per unit time of the process. For this process the probability of absorption was obtained by Bartlett (1946) and a summary of the derivation is given in the Appendix. The formula is

$$\begin{aligned} P(\delta|a',b',N) = 1-\Phi\{\sqrt{N}(b'-\delta)+a'/\sqrt{N}\} \\ +\Phi\{\sqrt{N}(b'-\delta)-a'/\sqrt{N}\}\exp\{2a'(\delta-b')\}, \end{aligned} \tag{6.4}$$

where $a' = a/\sigma$, $b' = b/\sigma$, and $\Phi(t)$ is the standard normal integral.

The approximations involved in this formula are the approximation of discrete $n$ by continuous $n$, and the effect of ignoring the lower boundary. Armitage obtains an upper bound for the latter effect, and he shows that for $b > 0$ this is negligible for the cases used.

Armitage suggests that in practice it might be more suitable to specify $\alpha$ and $N$, rather than $\alpha$ and $\beta$. He shows that, given $\alpha$ and $N$, $\delta$ from (6.3) is a single-valued monotonically decreasing function of $\beta$, with $\delta = \infty$ at $\beta = 0$, and $\delta = -\infty$ at $\beta = 1$. Thus equation (6.3) can be solved for the $\delta$ corresponding to set values of $(\alpha,\beta,N)$.

Further, the hypotheses $H_0$ and $H_1$, can be tested with the same errors $(2\alpha, \beta)$ by a fixed sample size plan, so that we can compare the fixed sample size $N_0$, with the maximum sample size on the 'restricted' plans. For $(2\alpha, \beta)$ in the range (0·01, 0·10) Armitage shows that $N/N_0$ varies from 1·25 to 1·55, and $N/N_0 = 1{\cdot}37$ if $2\alpha = \beta = 0{\cdot}05$. Thus the price paid for allowing for the possibility of stopping trials earlier is that, for all trials at which $H_0$ is accepted, between 25% and 55% more subjects will be needed than on the equivalent fixed sample size plan. Armitage later introduces a modification of the middle boundary to reduce the sample size for accepting $H_0$, see § 6.5.

## 6.4. Application to Tests for Binomial Probability

Suppose two alternative medical treatments are to be compared, and that they can both be given, in random order, to each patient. Suppose

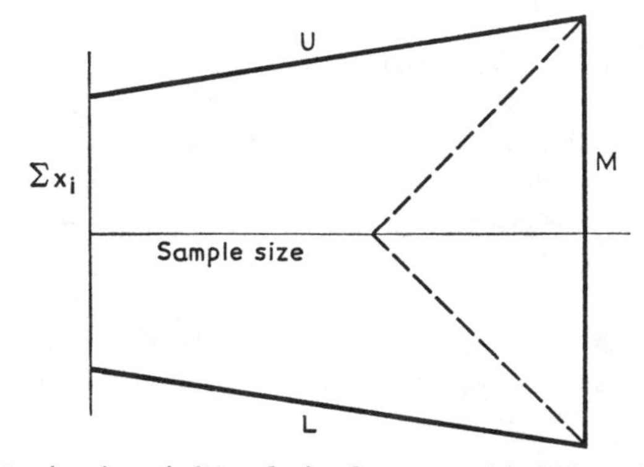


FIG. 6.1. Armitage's boundaries for a two-sided binomial test.

further that the result of a trial on a patient is a preference, denoted +1 or −1, for one or other treatment. Denote the probability that a patient gives a positive preference by $\theta$, then we require to design sequential trials for $\theta = \frac{1}{2}$ against two-sided alternatives. Open sequential designs for this problem were discussed in Chapter 3, but approximate closed designs can be based on the theory given above in § 6.2.

For large $n$, the sum $\sum x_i$ of the responses is nearly normally distributed with

$$\mu = 2\theta - 1 \quad \text{and} \quad \sigma^2 = 4\theta(1-\theta).$$

A straightforward application of the above theory leads to the boundaries U, L, M indicated in Fig. 6.1.

Now clearly the middle boundary M can be contracted as indicated by the dotted lines, for if the dotted lines are crossed, the path must eventually cross M.

Using the contracted form of boundaries, Armitage calculated some plans using the normal theory approximation, and some numerical checks of the accuracy showed remarkably good agreement. The ASN was calculated for one plan ($2\alpha = \beta = 0{\cdot}05$, $\theta_1 = 0{\cdot}8$), and although for this example the maximum ASN (using the contracted boundaries) was at most equal to the equivalent fixed sample size, very few observations are saved for $\theta$ in the range $(0{\cdot}35, 0{\cdot}65)$.

Bross (1952) obtained two sets of boundaries, rather similar to Armitage's, for a two-sided test of binomial $\theta$. The method of construction seems to have been largely trial and error.

## 6.5. Contraction of the Middle Boundary

We have noted that the outer boundaries of Armitage's restricted procedures coincide with the outer boundaries of a Sobel and Wald type of procedure. However, for the normal distribution case, the middle boundary of the restricted plan is very different from that given in § 3.2, being perpendicular instead of wedge shaped. Even with the contracted boundary obtained in the binomial case (Fig. 6.1), some further contraction of the middle boundary should be possible. Schneiderman and Armitage (1962a) suggested a method of contracting the middle boundaries of the restricted plan for the test of a normal mean and although the method is very arbitrary, it does appear to yield an improved ASN with very little change in the probabilities of error.

Any contraction of the middle boundary will reduce the probability $\alpha$ of reaching U when $\delta = 0$, and also reduce the probability $(1-\beta)$ of reaching U when $\delta = \delta_1$. This effect could be counteracted by picking a sample size $N' > N$ for U and L so that $\alpha' > \alpha$, and $\beta' < \beta$, but satisfying the relation

$$\frac{1-\beta'}{\alpha'} = \frac{1-\beta}{\alpha},$$

so that by (6.2), the straight lines for U and L are the same for $(\alpha', \beta')$ as for $(\alpha, \beta)$.

A contracted middle boundary M′ can now be found, which

reduces $\alpha'$ approximately to $\alpha$ again. This could be done in a large number of ways, but Schneiderman and Armitage propose choosing M′ so that the probability of a path from any point on M′ to U or L when $\delta = 0$, is a fixed quantity $2\epsilon$. Since the probability of reaching M′ when $\delta = 0$ is required to be $(1-2\alpha)$, the probability of reaching U or L when $\delta = 0$ is reduced by $2\epsilon(1-2\alpha)$ when M′ is used.

Therefore, when using outer boundaries U and L corresponding to $(\alpha', \beta', N')$, the use of the contracted boundary M′ alters the probability of reaching U or L when $\delta = 0$ to about $2\{\alpha' - \epsilon(1-2\alpha)\}$, and we require this to be equal to $\alpha$,

$$\alpha = \alpha' - \epsilon(1-2\alpha). \tag{6.5}$$

This equation gives the $\epsilon$ to be used corresponding to any chosen $N'$ or $\alpha'$ (or vice versa). Schneiderman and Armitage used $\epsilon = 0{\cdot}01\alpha$ in their plans, and this choice appeared to be arbitrary.

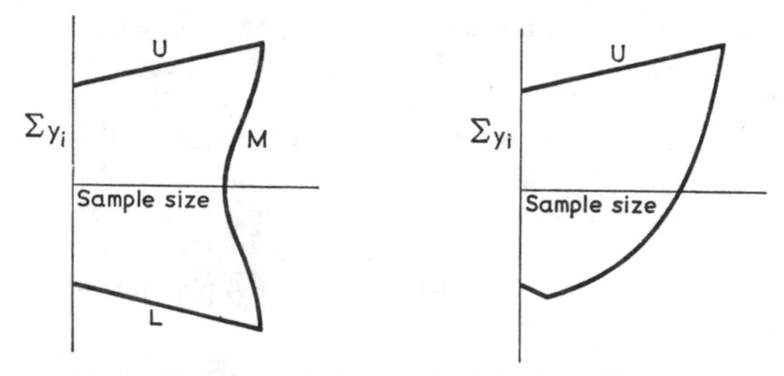


(a) 'Two-sided' boundaries (b) 'One-sided' boundaries

Fig. 6.2. Boundaries for one- and two-sided test of a normal mean.

It is difficult to examine the effect the contraction of the middle boundary has on $\beta'$, except that it probably brings it close to $\beta$ again.

Sets of boundaries for $(\alpha', \beta', N')$ can be obtained as in § 6.2, and the coordinates of the wedge-shaped middle boundary can also be obtained using the diffusion theory approximation.

We notice that in the calculation of the middle boundary, deletion of reference to L produces a one-sided test of $\mu = 0$ against $\mu > 0$. The shape of the boundaries is indicated in Fig. 6.2.

Some Monte Carlo runs were carried out as a check, and to estimate the ASN-curve for a one-sided and a two-sided plan. The results showed a considerable improvement in the ASN near $\mu = 0$.

### 6.6. Boundaries to Minimize the Maximum Expected Sample Size

Anderson (1960) considered tests of a normal mean $\mu$ when the variance $\sigma^2$ was known, using methods similar in principle to those used by Armitage (1957), but with much greater sophistication.

Suppose we wish to test $H_0 : \theta = \theta_0$, against $H_1 : \theta = \theta_1$, with equal probabilities of error $\alpha$. Define the sequential test in the plane

$$\left(n, \sum_{i=1}^{n} x_i\right)$$

where $x_i$ are the observations, supposed independent. Anderson considers tests having two straight-line convergent boundaries, with possible truncation at a maximum sample size $N$.

Following Armitage (1957), Anderson approximates the process in discrete $n$ by one in continuous $n$ as described in § 6.3, and the Appendix. Formulae are obtained for the probability of crossing either boundary first, before a sample size $n$, and for the expected sample size. It is conjectured that these formulae based on the continuous process are a good approximation to the discrete case, but no checks are given.

Anderson suggests that it might be reasonable to ask for the boundaries to have a minimum expected sample size at $\theta = \frac{1}{2}(\theta_1 + \theta_0)$. However, the formulae are too complicated for this to be done analytically, and two cases are examined numerically.

Anderson gives a more precise calculation of the operating characteristic than Armitage, using two boundaries, and the results can be generalized to include boundaries defined by a series of straight lines. If there were some further development, provision of tables, carrying out of checks, etc., Anderson's work would provide useful tests.

### 6.7. Discussion of Armitage's Methods

Armitage (1960) discusses the use of restricted procedures, Bross's plans, the sequential $t$-test, etc., in sequential medical trials. Anscombe (1963), in a review of Armitage's book, has criticized the whole formulation. Armitage (1963) gives a brief reply to Anscombe's arguments, and these two papers bring the main differences between two opposing schools of thought sharply into focus. One of the most important issues raised by Anscombe is a criticism of the whole theory of tests described so far in the present volume. Anscombe says:

> 'Very briefly, the objection to error probabilities of the two kinds is that these are expectations of something (of making a wrong decision), taken conditionally on the parameter values but *unconditionally over the whole sample space.* If such expectations are used as a means towards drawing inferences from some observations, the consequent inferences, beliefs and actions will perhaps be much affected by what was *not* observed, by all the rest of the sample space besides the one observed point in it. Absurdity can (and in the present case certainly will sometimes) result.'

Anscombe admits that, '... the concept of error probabilities of the first and second kinds, of power curve or operating characteristic, has some relevance to the design of impartial routine decision procedures, such as industrial inspection plans,' but he stresses '... it has no direct relevance to experimentation ...'.

These points are argued at length in Anscombe (1963) and Armitage (1963), to which the reader is referred. For related articles, see Barnard *et al.* (1962), Birnbaum (1962), and references. I summarize the discussions as follows.

Two aspects of a sequential procedure must be clearly distinguished, the stopping rule, and the manner in which inferences are made once observations are stopped. Although these two aspects are closely related, they involve a different set of concepts; in the design problem, it is important to know how probable are various possible results, and this involves considerations such as the OC-curve, probabilities of error, etc. Once the design problem is settled, then the likelihood, or posterior probabilities are mainly (if not alone) relevant. Likelihood statisticians argue – like Anscombe in the above quotations – that probabilities of error are not relevant to inference, but the force of these arguments is weakened a little if the method of accepting or rejecting the null hypothesis is not rigidly adhered to. In practice, the problem of making an inference is a complicated affair, involving much judgment, the consideration of side effects, and the need for further experimentation, etc., so that estimates of the effects would nearly always be made. Our conclusion is not merely, 'The null hypothesis is rejected'! Nevertheless, there is considerable force in the argument that probabilities of error are not relevant to inference.

The decision theory treatment of sequential designs is discussed in the next chapter (see also § 1.4), and the application to sequential medical trials is discussed by Anscombe (1963) and Colton (1963b). A Bayesian proceeds by defining a prior distribution for the unknown parameter, costs and losses, and determines boundaries which minimize the posterior risk. We can now point out that, while boundaries determined from a Neyman–Pearson standpoint may result in inferences depending on what was never observed, so the Bayesian boundaries may depend – through the prior distribution – on values of $\theta$ which never occur.

For the present problem a reasonable approach might be as follows. For the comparison of two alternative medical treatments, one treatment is often a standard, or one commonly used, and the other a new treatment. In such a case it is reasonable to require of any design – from an ethical standpoint – that it will not wrongly accept the new treatment with a probability greater than $\alpha$. Having fixed this, it is reasonable to fix the boundaries by minimizing some overall utility (see Horsnell (1957) for an approach along these lines).

Another line of development has been pursued by Armitage (1967), Armitage *et al.* (1969), and McPherson and Armitage (1971). This begins with a consideration of the 'optional stopping effect', which can be illustrated as follows. Suppose we have a series of independent normally distributed observations $x_1, x_2, \ldots$, which measure the difference between two treatments. The null hypothesis is that the mean $\mu = 0$, with a known variance $\sigma^2$. It is clear that a reasonable stopping rule from a likelihood or Bayesian viewpoint is to stop as soon as

$$\left| \sum_1^n x_i \right| > k\,\sigma\sqrt{n}. \tag{6.6}$$

However, it is well known that if $\mu = 0$, this stopping boundary must be reached eventually. Even if $\mu$ is slightly negative, there is a substantial probability of reaching a decision that $\mu > 0$, by termination under (6.6). For example, if $k = 1{\cdot}96$, the probability of termination under (6.6) is $0{\cdot}266$ by $n = 25$ and $0{\cdot}374$ by $n = 100$. [It should be pointed out that likelihood or Bayesian statisticians may wriggle out of the dilemma by insisting on presenting the likelihood or posterior distribution of $\mu$ instead of 'reaching a decision'.]

Armitage's solution is as follows. Suppose we require a test that a normal mean $\mu=0$ versus two-sided alternatives. We can use outer boundaries, at which the alternative hypotheses are accepted, of the form (6.6), instead of the straight line boundaries (6.1). The experiment is terminated with acceptance of the null hypothesis at a maximum sample size $N$, which is small enough to make the accumulated probability of reaching the outer boundaries under the null hypothesis fairly small. This proposal produces boundaries very similar to those discussed in Section 6.1, except that the outer boundaries are nearly parabolic instead of linear. Similar principles apply to tests of binomial probability, and McPherson and Armitage (1971) contains sequential plans for both binomial and normal variables.

It is possible that these latest proposals by Armitage may be more acceptable to likelihood and Bayesian statisticians than the ones given earlier, and the question requires further study.

The references to this chapter list some of the published applications of Armitage or Bross closed sequential plans (Sequential medical trials using an SPRT or sequential $t$-test design are not listed.) I am indebted to Professor P. Armitage for some of the references given.

**Problems 6**

1. Outline a series of empirical sampling trials for an electronic computer, in order
(a) to study the properties of the plans discussed in this chapter;
(b) to compare the plans in this chapter with the SPRT plans discussed in Chapters 2 and 3.

Explain how you would present your results. (See **P4.6 and P5.4.**)

2. The problems in this volume have suggested extensive use of empirical sampling trials for the study of sequential plans. An alternative method would be to programme a computer to work successively from one sample size to the next, using numerical integration, and tabulating the distribution of sample size, the distribution of overflow of the boundaries, the probabilities of error, etc. With closed plans this method may be preferred to empirical sampling. Outline a flow chart for a programme to study the properties of Armitage's two-sided test of a normal mean.

3. If estimates – not merely decisions – are required from the results

of sequential trials, many statisticians would base such estimates on the overall sampling distribution of results over the boundaries. While none of the plans given so far in this book has been designed for estimation, in many practical cases estimation will be a subsidiary aim. Consider one-sided tests of a normal mean, $\sigma^2$ known, and outline a method of comparing the sampling distributions of the sample mean of Armitage's plan and the SPRT. What results do you expect to obtain and which plan do you think will be better from the estimation point of view?

4. Cox (1963), in the discussion of his asymptotic maximum likelihood test procedure (see § 4.7), said that closed tests could be obtained following Armitage's methods. Obtain a closed sequential $t$-test, and outline a study comparing its properties with the open sequential $t$-tests.

# CHAPTER 7

# Decision Theory

## 7.1. Assumptions

There is now a large literature on decision theory, concerning both the mathematics of specific applications, and the underlying philosophy of various approaches to the subject. In this chapter we give a brief summary of some of the aspects more important to the derivation of sequential statistical procedures. It will be convenient to use the sampling inspection illustration used in § 1.4; that is, we shall consider the determination of sequential plans to accept or reject large batches of items. In § 1.4 we defined a cost of sampling, costs of making wrong decisions, and a prior distribution; unfortunately, knowledge of these quantities is usually vague, and it is essential to realize the implications of specific assumptions.

The terminology used in this chapter is as follows. The functions defining the costs of wrong decisions are decision loss functions. For any plan, and for a given value of $\theta$, the sum of all decision losses and sampling costs, averaged over the set of possible outcomes, is called a risk.

*Cost of inspection*

In most applications the cost of inspection has been taken to be proportional to the number of items inspected, and with this assumption a term proportional to the ASN-function appears in the loss function, see equation (1.4). In fact equation (1.4) is merely a weighted combination of the OC-curve and ASN-function, and its purpose can be regarded as giving some overall loss of a plan, as a function of $\theta$. This loss function has relative value only, in comparison with the loss functions of other plans. Therefore the assumption that the cost of inspection is a linear function of the number of items inspected means that in determining a sampling plan to use, the sample size distribution enters only via the ASN-function. More particularly we imply

(a) that however large a sample size happens to be, the cost is still

proportional to $n$, and there is no cost over and above this due to the hold-up in the flow of batches caused by a large sample size; and
(b) that variations in the inspection load are acceptable at no extra cost.

These assumptions are also made in the SPRT formulation, of course, since the only way the cost of sampling enters is by consideration of the ASN-function. In practice both very large sample sizes and also large variations in sample size are often regarded as unacceptable. This suggests a different approach is necessary, and Armitage (1960) and Bross (1952) started afresh to generate *ad hoc* procedures with more suitable properties; see Chapter 6 for details. With the decision theory formulation, extra terms can be added to the cost function, such as

$$\text{cost of inspecting } n \text{ items} = an+bn^2,$$

etc. A detailed consideration of suitable forms for the cost function on the basis of case histories is overdue.

*Cost of wrong decisions*

In some of the illustrations given in this book the costs of wrong decisions are easy to assess in monetary terms. For example, consider the problem of acceptance sampling of electronic components for a computer, Ex. 1.3. For another example, consider the forest insect survey, Ex. 3.2, where the cost of needless spraying is known, and the loss due to omission of spraying can presumably be estimated. However, in the whitefish sampling problem, the cost of passing a bad car load of fish lies in loss of prestige, etc., and is difficult to estimate.

In many situations monetary cost considerations are inappropriate, but utilities can be defined and are sometimes easy to assess and more meaningful than probabilities of error. Anscombe (1963), for instance, has argued at length that when designing sequential medical trials a formulation in terms of utilities should be used in preference to Armitage's methods.

There is one class of situation where probabilities of error are to be preferred to utilities (or used in addition), and for an illustration consider the sequential grading of milk example, Ex. 2.3. Here, some public authority may justifiably require limits on the probability of milk of a certain grade being passed for sale. There are a number of common inspection problems which come under this heading, such as British Standards Institution requirements on electrical insulation.

However, in most situations the utilities of various possible decisions are relevant, and the only difficulty which arises is whether they can be estimated. If two terminal decisions are being made, there is a break-even quality, at which it is just as costly to take either decision, and it has been shown that plans are more sensitive to the break-even quality assumed than to the exact form of decision losses.

*Ex. 7.1.* A possible form for the loss functions of Ex. 1.3 is as follows. Let the cost of inspecting an item (assumed constant) be the unit of costs, and let

$$W_0(\theta) = \begin{cases} k_0(\theta-\theta_0) & (\theta \geqslant \theta_0) \\ 0 & (\theta \leqslant \theta_0) \end{cases}$$

and

$$W_1(\theta) = \begin{cases} k_1(\theta_0-\theta) & (\theta \leqslant \theta_0) \\ 0 & (\theta \geqslant \theta_0), \end{cases} \tag{7.1}$$

where $\theta_0$ is the break-even quality. Equation (1.4) becomes

$$R(\theta|S) = \begin{cases} E(n|\theta,S)+k_0 P_A(\theta|S).(\theta-\theta_0) & (\theta \geqslant \theta_0) \\ E(n|\theta,S)+k_1 P_R(\theta|S).(\theta_0-\theta) & (\theta \leqslant \theta_0) \end{cases} \tag{7.2}$$

where $S$ specifies the sampling plan used.

A plan $S$ is preferred if it has a lower $R(\theta|S)$ than other plans, and usually different plans will be preferred in different ranges of $\theta$. A plan such that there is no other plan preferred uniformly in $\theta$ is said to be admissible. In § 1.4 two main ways of selecting a single plan from the class of all admissible plans were mentioned. Since the parameters of the loss function, such as $k_0$, $k_1$, $\theta_0$, in Ex. 7.1 above, are usually known only approximately, it is important to know what effect variations in these parameters will have, and this will clearly depend on how the particular plan is chosen.

If $k_0$ and $k_1$ are both multiplied by the same constant, this results in wrong decisions costing more, and the final plan will usually have an increased sample size, and lower probabilities of error. On the other hand, variations in the ratio of $k_0$ to $k_1$, or in $\theta_0$, will (usually) lead to an increase in one of the probabilities of error and a decrease in the other. Thus doubt about the costs of wrong decisions is paralleled by the arbitrary way in which the probabilities of error are assigned in formulations discussed in earlier chapters.

*The prior distribution*

Use of the risk function $R(\theta|S)$ to choose a sampling plan is a means of balancing the costs of sampling against the costs of wrong decisions. Consideration of $R(\theta|S)$ alone never leads to a unique plan, and it is reasonable to choose a plan which operates best in the region of $\theta$ most commonly used, or which operates best in some average sense, over likely values of $\theta$. If information is available on the long-run relative frequency with which values of $\theta$ occur (called the prior distribution), this should be considered when determining a plan. In some applications, back inspection records, etc., will be available from which we may form an estimate of the prior distribution. Usually such records will be of samples from batches, so that the recorded data are sampled from a population distribution which is the convolution of the prior distribution and the distribution of sample results from a given batch. In the deconvolution process there may be considerable latitude on the models to be used. However, insufficient information on the prior distribution should not prevent its being used, provided we can check that the final plan is insensitive to likely variations in the prior.

In other applications there may be no information about the prior distribution, in which case its use could be avoided by using the minimax principle (see § 1.4 and § 7.7), or else we could argue as follows. Wald has shown that, under very weak conditions, the class of all admissible plans is generated by minimizing the risk with respect to all possible prior distributions. Other authors argue that a few principles of consistent behaviour indicate that a person should act as if he had a prior distribution. In this sense, it is reasonable to generate a set of plans by using prior distributions, but index and use them by other criteria.

One argument against prior distributions is that even if they were fully determined, the operation of a sampling plan will often change the prior distribution. (Indeed, sometimes this is the aim of sampling plans, see Hill, 1960.) However, the prior distribution could be periodically updated, using sample data. Alternatively a family of prior distributions could be assumed, and after each completed sample, the prior distribution itself could be changed by use of Bayes's Theorem. An important method involving estimation of the prior distribution from data, called 'Empirical Bayes', is discussed in § 7.6.

*General comments*

In the above discussion it has been implied that batches are being sentenced independently. Anscombe (1961) has pointed out that for a consumer, what is predetermined is not the quantity inspected, but the quantity bought, and in this case there is an obvious interdependence between inspection decisions.

Some discussion of the formulation of costs and losses appears in Anscombe (1961), Hald (1960), and there are brief comments in Anscombe (1958). Anscombe (1961) contains some discussion of how costs and losses might work out for the inspection of electricity meters. Horsnell (1957) formulates a model involving costs, which is quite different from that used here, and the costs of wrong decisions are not explicitly introduced; see P7.5.

Most of the arguments against the decision theory approach rest on the uncertainty of knowledge of both form and value of the losses and of the prior distribution. Some studies by Hald (1965), and Wetherill and Campling (1966) (see also Wetherill, 1960, and Pfanzagl, 1963), have shown, however, that the optimum is very robust to practically everything except the break-even quality, $\theta_0$, which turns out to be critically important. This indicates that practical considerations should include economic studies directed at estimating the break-even quality. The fact that Bayes solutions seem to be reasonably robust to the prior distribution and the numerical values of the decision losses (other than $\theta_0$) largely answers the objections referred to.

In any case, variations in the parameters of the decision theory model are reflected by parallel uncertainties in the formulations described earlier. By attempting to formalize the choice of a plan in terms of decision theory, the important implications of assuming specific forms for the cost of inspection, etc., readily come to light. By considering probabilities of error, and considering only the ASN-function, we camouflage the assumptions really being made. When designing a sequential test, it is often instructive to attempt a decision theory formulation, even if the result is not formally used. See Griffiths and Rao (1964) for an application of decision theory to a sampling inspection problem, which led to a much tighter plan being adopted than one selected previously.

Probabilities of error have been used so commonly by statisticians, that there is the feeling that in this metric at least we have real, meaningful quantities. However, in many applications – such as those referred to in this volume – they are not directly relevant and sometimes meaningless (consider Exs. 2.2, 3.2). Although the scale for costs is arbitrary in most cases this does represent more nearly a natural metric for comparison of plans.

### 7.2. Bayesian Approach

The foundation of this approach is Bayes's theorem, which we develop as follows. Suppose $E$ and $H$ are any two events which can occur together, then by the rule for conditional probabilities,

$$\Pr\{H|E\} = \Pr\{EH\}/\Pr\{E\}. \tag{7.3}$$

Now consider a set of events $H = \{H_1, H_2, \ldots, H_k\}$, one and only one of which must occur at a trial, then

$$\Pr\{E\} = \sum_{j=1}^{k} \Pr\{E|H_j\}\Pr\{H_j\}. \tag{7.4}$$

By inserting (7.4) in (7.3) we obtain

$$\Pr\{H_j|E\} = \frac{\Pr\{E|H_j\}\Pr\{H_j\}}{\sum_{l=1}^{k} \Pr\{E|H_l\}\Pr\{H_l\}}. \tag{7.5}$$

This is an exact relation between conditional probabilities called Bayes's theorem. The interpretation of (7.5) is as follows.

Consider a sequence of events at which one and only one of $H_1, \ldots, H_k$, must occur, and also one and and only one of the events $A, B, C, D, E, \ldots$, must occur, so that we obtain a sequence

$$(H_2, D), (H_1, E), (H_7, A), \ldots$$

Out of this sequence select the events in which $E$ occurs,

$$(H_1, E), (H_4, E), (H_1, E), (H_2, E), \ldots,$$

then equation (7.5) gives the limiting relative frequency of $H_j$ in this subsequence. Clearly, for this to have much meaning, all events $H_j$ should be statistically independent (and also separately, the events $A, B, \ldots$).

The final step here is to consider the $H_j$ as hypotheses, the $\Pr\{H_j\}$ as prior probabilities, then equation (7.5) gives the posterior probabilities of $H_j$, given an observed event $E$.

Barnard (1954) showed that Bayes's theorem can still be applied (in terms of frequencies), even if the $H_j$ are not a random sequence. See the source paper for details.

*Ex. 7.2.* Suppose large batches of items are presented for acceptance inspection, in which the proportion of bad items is known to be either $\theta_0$ or $\theta_1$, with prior probabilities $a$ and $(1-a)$ respectively. After $n$ items are inspected from a batch, $r$ are found defective. The posterior probability of $\theta_0$ is, by (7.5)

$$\frac{a\theta_0^r(1-\theta_0)^{n-r}}{a\theta_0^r(1-\theta_0)^{n-r}+(1-a)\,\theta_1^r(1-\theta_1)^{n-r}}. \tag{7.6}$$

(The posterior probability of $\theta_1$ is one minus this.)

To fit this into the above interpretation, we put $H_1:\theta=\theta_0$, $H_2:\theta=\theta_1$, and we suppose that these occur with probabilities $a$ and $(1-a)$ respectively. The events $A$, $B$, $C$, $D$, ..., are the set of all possible results from inspecting a batch. From the combined set $(H_i, A)$, etc., defined, we select a subsequence in which the observed result $(n, r)$ occurred, and in this subsequence, the limiting relative frequency of $\theta_0$ batches is given by (7.6). That is, expression (7.6) is the conditional probability that the batch being observed has quality $\theta_0$ given the observed results; the conditional probability is dependent upon the assumed prior distribution, although this effect reduces as the sample size increases.

*Ex. 7.3.* Suppose random variables $x_1, x_2, \ldots, x_n$ are observed from a logistic distribution with a p.d.f.

$$\frac{\exp(x-\theta)}{\{1+\exp(x-\theta)\}^2}$$

where the prior distribution of $\theta$ is standard normal. The posterior distribution of $\theta$ is

$$\frac{\prod_{i=1}^{n}\left\{1+\exp(x_i-\theta)\right\}^{-2}\exp(-n\theta-\tfrac{1}{2}\theta^2)}{\int_{-\infty}^{\infty}\prod_{l=1}^{n}\left\{1+\exp(x_l-\theta)\right\}^{-2}\exp(-n\theta-\tfrac{1}{2}\theta^2)\,\mathrm{d}\theta}.$$

This last example illustrates an important point. Often posterior distributions increase in complexity with increasing $n$, and when this happens it is extremely tedious to use a Bayesian approach in a complex sequential analysis set-up, and there is very little hope of obtaining any theoretical results. It is usual, therefore, to consider a restricted class of distributions called distributions closed under sampling, in which the posterior distribution not only remains simple, but also remains of the same form as the prior distribution.

*Conjugate prior distributions*

Let the probability density of observations $x$ be $\phi(x|\theta)$, where $\theta$ is an unknown parameter, and suppose the prior distribution for $\theta$ is of the form $\xi(\theta|\alpha)$, where the parameter $\alpha$ indexes a unique distribution from the family $\xi$. Then the family of prior distributions $\xi(\theta|\alpha)$ is said to be a conjugate of the distribution $\phi(x|\theta)$ if for all $x_1, x_2, \ldots, x_n$, for which $\phi(x_1|\theta)\neq 0$, there exists a $\beta=\beta(\alpha, x_1, \ldots, x_n)$ such that

$$\xi(\theta|\beta) = \frac{\xi(\theta|\alpha)\,\phi(x_1|\theta)\ldots\phi(x_n|\theta)}{\int \xi(\theta|\alpha)\,\phi(x_1|\theta)\ldots\phi(x_n|\theta)\mathrm{d}\theta}. \tag{7.7}$$

*Ex. 7.4.* Denote the probability density function of the observations by $\phi(x|\theta)$, where the $\theta$ can take on certain values $\theta_1, \ldots, \theta_k$, and let the prior probability of $\theta_i$ be $a_i \geqslant 0$, where $a_1+\ldots+a_k = 1$. The prior distribution is $(a_1,\ldots,a_k)$, and after observations $x_1, x_2, \ldots, x_n$, the posterior distribution is of the same form $(b_1,\ldots,b_n)$, where

$$b_j = a_j \prod_{i=1}^{n} \phi(x_i|\theta_j) \Big/ \sum_{l=1}^{k} a_l \prod_{r=1}^{n} \phi(x_r|\theta_l).$$

*Ex. 7.5.* Suppose that observations $x_i$ have a normal distribution with mean $m$ and unit variance, denoted $N(m, 1)$, and the prior distribution for $m$ is $N(\mu, \sigma^2)$. After observations $x_1, \ldots, x_n$, the posterior distribution of $m$ is

$$N\left(\frac{n\sigma^2\bar{x}_n+\mu}{n\sigma^2+1}, \frac{\sigma^2}{1+n\sigma^2}\right),$$

where $$\bar{x}_n = \frac{1}{n}\sum_{1}^{n} x_i.$$

A very important property of conjugate prior distributions was proved by Stone.

*Theorem 7.1* (*M. Stone*). For continuous prior distributions the posterior distribution is of the form

$$K\{G(\theta)\}^{\beta_1} \exp\{\mathrm{A}(\theta)\beta_2\}$$

where $\beta_1$ is the sample size plus an arbitrary constant and $\beta_2$ depends on the observations.

*Proof*. Now $\beta$ of (7.7) must be a sufficient statistic, and the only distributions possessing a sufficient statistic of fixed dimension for all sample sizes are those of the exponential family. Therefore the distribution of observations $x_i$ must be of the form

$$F(x) . G(\theta) \exp \{\mathrm{A}(\theta)\beta_2\}$$

Suppose there is a continuous prior distribution $\xi(\theta|\alpha)$, then the posterior distribution after observations $x_1, \ldots, x_n$ is

$$\frac{\xi(\theta|\alpha)\, \{G(\theta)\}^n . \mathrm{e}^{A(\theta)\Sigma B(x_i)}}{\int \xi(\theta|\alpha)\, \{G(\theta)\}^n . \mathrm{e}^{A(\theta)\Sigma B(x^i)}\, \mathrm{d}\theta}$$

since the $F(x_i)$ terms cancel. A continuous prior distribution must therefore be of the form

$$K\{G(\theta)\}^{\beta_1} \exp \{A(\theta)\beta_2\}, \tag{7.8}$$

where $\beta_1$ is the sample size $n$ plus an arbitrary constant. There is an obvious generalization of this theorem when $\theta$ is a vector.

The importance of the theorem can be explained as follows. Suppose we have a conjugate family of prior distributions $\xi(\theta|\alpha)$, where $\alpha = (\alpha_1, \alpha_2, \ldots, \alpha_t)$, then the effect of sampling can be described by a random walk in a $\xi$-space, in which the dimensions are components $\alpha_1, \alpha_2, \ldots, \alpha_t$ of $\alpha$. Stone's theorem shows that this walk is restricted, and that for increasing sample size, no prior distribution can repeat itself.

There are exceptions to this. If $\xi(\theta)$ is taken to be discrete, as in Ex. 7.4, then it is possible under certain circumstances for a prior distribution to repeat itself. In Ex. 7.2, for instance, there exists an infinite set of $n, r$ for which

$$\theta_0^r(1-\theta_0)^{n-r} = \theta_1^r(1-\theta_1)^{n-r},$$

and the prior distribution repeats itself. However, usually the sample size is a dimension of the $\xi$-space.

## 7.3. Sequential Sampling

A sequential plan is said to be a Bayes solution if it is obtained by minimizing some overall risk averaged over a prior distribution. This section discusses the Bayes solution for a simple acceptance sampling problem, when the prior distribution is limited to distributions closed under sampling. Wetherill (1961) has given a more complete discussion of a number of points arising in this section and § 7.4.

Suppose observations $x_i$ are taken on (large) batches, and the observations are independently distributed with a probability density function $\phi(x|\theta)$, and the prior distribution of $\theta$ is $\xi(\theta|\alpha)$, which is a conjugate distribution. As a result of inspection one of two terminal decisions is to be made, which we label decisions 1 and 2. The loss of taking decision $i$ when the true quality of the batch is given by $\theta$ is $W_i(\theta)$. The cost of inspecting an item (supposed constant) is taken as the unit of costs.

For any fixed sample size $n$, the risk in taking decision '$i$', conditionally upon observations $x_1, \ldots, x_n$, is

$$D_i(\alpha, x_1, \ldots, x_n) = \frac{\int \xi(\theta|\alpha)\, \phi(x_1|\theta) \ldots \phi(x_n|\theta)\, W_i(\theta)\, \mathrm{d}\theta}{\int \xi(\theta|\alpha)\, \phi(x_1|\theta) \ldots \phi(x_n|\theta)\, \mathrm{d}\theta}. \quad (7.9)$$

By (7.7) this is equal to

$$D_i(\beta) = \int \xi(\theta|\beta)\, W_i(\theta)\, \mathrm{d}\theta,$$

and decision 1 is preferred to decision 2 if $D_1(\beta) < D_2(\beta)$. Thus we have Theorem 7.2.

*Theorem 7.2.* The terminal decision to be made depends only on the posterior distribution of $\theta$.

Therefore, all points in $\xi$-space can be classified as terminal decision 1 or 2 points. Corresponding to any set of observations, there is a single point in the $\xi$-space, although several different samples $(x_1, \ldots, x_n)$ may correspond to the same single point in $\xi$-space. Stone's theorem shows that in general, where several samples lead to the same point in $\xi$-space, they must be of the same size.

In general any sequential sampling plan is represented by a division of the $\xi$-space into three regions, two terminal decision regions and a continuation region. This result is almost obvious, and rests on the fact that the continuation risk at any point $\alpha_n$ in $\xi$-space (corresponding to sample size $n$), depends only on $\alpha_n$ and the decision boundaries for samples of size greater than $n$, and it does not depend on the particular sample path taken to reach $\alpha_n$.

At any point $\beta$ in the $\xi$-space we can define the continuation risk $C(\beta)$, and the terminal decision risks $D_1(\beta)$, $D_2(\beta)$. The point $\beta$ is a continuation point, or a decision 1 or 2 point respectively according to which of these three risks is the least. The Bayesian solution is completely described by the boundaries between these various regions. This brings us to the basic difficulty of this approach, that in order to define the continuation risk at $\beta$, it is necessary to know the decision boundaries for all points leading from $\beta$. There are several ways around this difficulty.

(i) The decision boundaries may meet, in which case the decision boundaries can be worked out backwards from this point. The next section describes how the meeting point may be determined.

(ii) An approximate solution can be obtained by forcing the boundaries to meet at a very large sample size.

(iii) Even when the decision boundaries do not meet, it is still possible in certain cases to write down the equations they satisfy, and solve them (see Wetherill, 1959; Vagholkar and Wetherill, 1960).

In this discussion of sequential sampling we have assumed the cost of sampling to be independent of the serial number $n$ of the item being sampled. If the cost of sampling is a function of $n$, it remains true that the Bayes solution is a division of the $\xi$-space into three regions, provided $n$ is a coordinate of the $\xi$-space.

Stone's theorem indicates that generally $n$ will be one of the coordinates of the $\xi$-space, but there are exceptions. If a prior distribution can repeat itself, and if the cost of sampling is a function of $n$, then a dimension for $n$ must be added to the $\xi$-space to obtain the space in which the Bayes solution is defined.

*Ex. 7.6.* The $\xi$-space for the two point binomial prior distribution of Ex. 7.2 is a single parameter taking values in the open interval $(0,1)$. If the cost of sampling depends on $n$, the Bayes solution is defined on $(a,n)$.

### 7.4. The Meeting Point

It follows from Stone's theorem that the $\xi$-space can be specified by $(n,y)$, where $y$ is a sufficient statistic, supposed continuous. Under reasonably mild conditions on the loss functions, there will be, for each $n$, a value of $y$, $y_0(n)$ say, for which decision 1 is preferred for all $y < y_0(n)$, and decision 2 is preferred for all $y > y_0(n)$. The locus

$(n, y_0(n))$ is called the neutral boundary. The decision 1 and 2 boundaries will be two values of $y$, say $y_1(n)$ and $y_2(n)$, so that

$$y_1(n) \leqslant y_0(n) \leqslant y_2(n)$$

and the range $(y_1(n), y_2(n))$ is the continuation set at the sample size $n$.

Now it is easy to show that if one decision boundary meets the neutral boundary, then the other decision boundary does so at the same point. Suppose all three boundaries meet at a point $(N, y_0(N))$, then at this point:

Risk of making either terminal decision

= Risk of making one more observation and then making the relevant terminal decision.

This equation provides us with a means of determining the meeting point boundary $(N, y_0(N))$.

The precise conditions under which a meeting point boundary exists are not known. However, with continuous prior distributions, and some very mild restrictions on the cost of sampling and loss functions, a meeting point boundary exists. (Continuous prior distributions are not necessary. See Ray (1965) for work relevant to this point.)

The points $((N-1), y)$ all lead to points in $(N, y)$, the risks of which are now established. Thus the risk of taking either terminal decision, and the continuation risk can be calculated at all points $((N-1), y)$ and the boundaries $(y_1(N-1), y_2(N-1))$ determined. In this way we can work back and classify all points in the space $(n, y)$ as terminal decision or continuation points; see the examples below.

*Ex. 7.7.* Suppose we have the two point prior distribution of Ex. 7.2, and we wish to make one of two terminal decisions, which we label 1 and 2. The loss of taking decision $i$ when $\theta_j$ is true is written $W_{ij}$ where $W_{10} = W_{21} = 0$, and the cost of sampling the first item from a batch is taken as the unit of costs. Let the cost of sampling the $n$th unit be $\{1+f(n)\}$.

The neutral line is

$$W_{20} a \theta_0^r (1-\theta_0)^{n-r} = W_{11}(1-a)\, \theta_1^r (1-\theta_1)^{n-r} \tag{7.10}$$

or

$$W_{11}/W_{20} = a'/(1-a') \tag{7.11}$$

where $a'$ is the posterior probability of $\theta_0$,

$$a' = \frac{a\theta_0^r(1-\theta_0)^{n-r}}{a\theta_0^r(1-\theta_0)^{n-r}+(1-a)\,\theta_1^r(1-\theta_0)^{n-r}}.$$

Suppose a meeting point exists at a sample size $N$, and one failure (which has a probability $\theta$), leads to terminal decision 2 and one success leads to terminal decision 1. At $N$, the risk of taking one more observation is

$$1+f(N)+(1-a')\,(1-\theta_1)\,W_{11}+a'\,\theta_0\,W_{20}. \tag{7.12}$$

The risk of making a terminal decision at the meeting point is equal to the cost of a randomized decision at this point, with probability $\frac{1}{2}$ of decision 1 and 2, which is

$$\tfrac{1}{2}\{a'\,W_{20}+(1-a')\,W_{11}\}. \tag{7.13}$$

At the meeting point equation (7.12) must equal (7.13) so that

$$1+f(N)+(1-a')\,(1-\theta_1)\,W_{11}+a'\,\theta_0\,W_{20} = \tfrac{1}{2}\{a'\,W_{20}+(1-a')\,W_{11}\},$$

or $$\tfrac{1}{2}\{(1-a')\,(2\theta_1-1)\,W_{11}+a'(1-2\theta_0)\,W_{20}\}-1 = f(N). \tag{7.14}$$

Now if $f(n)$ is zero, the optimum plan is an SPRT, and it is easy to show that the expression on the left of (7.14) is zero or less only when there is no continuation region, see problem P7.9.

If $f(N)$ is not constant a meeting point will exist, in general. Suppose $f(N) = N^{\gamma}-1$, then by substituting this and (7.11) in (7.14) we obtain

$$N^{\gamma} = \frac{W_{11}\,W_{20}}{(W_{11}+W_{20})}\,(\theta_1-\theta_0),$$

and $N \to \infty$ as $\gamma \to 0$. When $\gamma = 0$, the decision boundaries are parallel, and the Bayes solution is the SPRT.

*Ex. 7.8.* Suppose that acceptance inspection of large batches of items is being carried out, and items are classified as defective or effective. Let the probability of a defective item be $\theta$, and the prior distribution for $\theta$

$$\theta^{s-1}(1-\theta)^{t-1}/\beta(s,t).$$

Suppose the loss of accepting a batch is

$$W_1 = \begin{cases} k(\theta-\theta_0) & (\theta \geqslant \theta_0) \\ 0 & (\theta \leqslant \theta_0), \end{cases}$$

and that the loss of rejecting a batch is

$$W_2 = \begin{cases} k(\theta_0 - \theta) & (\theta \leqslant \theta_0) \\ 0 & (\theta \geqslant \theta_0). \end{cases}$$

Denote the number of good and bad items found during inspection by $x$ and $y$ respectively, then the neutral line is

$$(y+s)(1-\theta_0)/\theta_0 = x+t$$

and that the meeting point is at $(X, Y)$, where

$$\left.\begin{aligned} X &= (Y+s)(1-\theta_0)/\theta_0 - t \\ Y &= k\theta_0^2(1-\theta_0) - \theta_0 - s \end{aligned}\right\}.$$

It is appropriate to make some brief comments here on Ex. 7.7 and Ex. 7.8. I conjecture that, under very general conditions on the loss functions, the only assumptions for which the binomial SPRT is optimum (in the decision theory sense) is the two point prior distribution, and linear cost function. Thus from a decision theory point of view, the SPRT is optimum only under a rather unlikely set of assumptions.

*Ex. 7.9.* Suppose we have the set-up described in Ex. 7.5, and that there are two terminal decisions, $D_1$ and $D_2$ to be made, with loss functions

$$W_1 = \begin{cases} k(m - m_0) & (m > m_0) \\ 0 & (m < m_0), \end{cases}$$

and

$$W_2 = \begin{cases} k(m_0 - m) & (m < m_0) \\ 0 & (m > m_0), \end{cases}$$

where the cost of one observation, supposed constant, is the unit of costs.

Let the posterior distribution for $m$ be written $N(\mu_n, \sigma_n^2)$, so that Ex. 7.5 states that

$$\begin{aligned} \mu_n &= (n\sigma^2 \bar{x}_n + \mu)/(n\sigma^2 + 1) \\ \sigma_n^2 &= \sigma^2/(1 + n\sigma^2). \end{aligned} \tag{7.15}$$

Clearly, any sample point is specified completely by $(\mu_n, n)$.

The risk of taking terminal decision $D_1$ is

$$D_1(\mu_n, n) = k \int_{m_0}^{\infty} \frac{(m-m_0)}{\sqrt{(2\pi)}\,\sigma_n} \exp\left\{-\frac{(m-\mu_n)^2}{2\sigma_n^2}\right\} dm$$

and for $D_2$

$$D_2(\mu_n, n) = k \int_{-\infty}^{m_0} \frac{(m_0-m)}{\sqrt{(2\pi)}\,\sigma_n} \exp\left\{-\frac{(m-\mu_n)^2}{2\sigma_n^2}\right\} dm.$$

The neutral boundary is obtained by equating $D_1(\mu_n, n)$ and $D_2(\mu_n, n)$, which yields $\mu_n = m_0$, for all $n$.

The continuation risk at $(\mu_n, n)$ can be written $C(\mu_n, n)$, and this is the expected cost of sampling from this point plus the expected cost of wrong decisions. Any point $(\mu_n, n)$ is classified as a continuation point, or as a terminal decision $D_1$ or $D_2$ point, according to which risk minimizes

$$R(\mu_n, n) = \text{Min}\{D_1(\mu_n, n), C(\mu_n, n), D_2(\mu_n, n)\}.$$

In order to determine the boundaries at a sample size $n$, the boundaries for all sample sizes greater than this must be known, so that $C(\mu_n, n)$ can be calculated. However, if a meeting point exists at a sample size $N$, all points corresponding to $(N+1)$ are decision points, and their risks are known.

The meeting point must be on the neutral boundary, and the risk at this point must be equal† to the risk of taking one more observation only, and then making a terminal decision. Further, if we are exactly at $(m_0, N)$, one more observation leads to a $D_2$ or $D_1$ point according to whether it is greater or less than $m_0$.

If we are at the point $(m_0, N)$, and the next observation is $x$, the posterior distribution at $(N+1)$ has mean and variance

$$\mu'_{N+1} = \{\sigma^2(Nm_0+x)+m_0\}/\{(N+1)\,\sigma^2+1\},$$

and

$$\sigma_{N+1}^2 = \sigma^2/\{1+(N+1)\,\sigma^2\}.$$

† Strictly there is an inequality involved here since $N$ is restricted to be integral. In order to simplify the discussion here, I will ignore this point.

The continuation risk from $(m_0, N)$ is therefore†

$$C(m_0, N) = 1 + \int_{-\infty}^{\infty} \left\{ \int_{-\infty}^{\infty} \frac{1}{\sqrt{(2\pi)}} \exp\left\{-\frac{(x-m)^2}{2}\right\} \frac{1}{\sqrt{(2\pi)}\,\sigma_N} \right.$$

$$\left. \exp\left\{-\frac{(m-m_0)^2}{2\sigma_N^2}\right\} \mathrm{d}m \right\} . R(\mu'_{N+1}, N+1)\,\mathrm{d}x, \qquad (7.16)$$

where

$$R(\mu'_{N+1}, N+1) = \mathrm{Min}\{D_1(\mu'_{N+1}, N+1), D_2(\mu'_{N+1}, N+1)\}.$$

Therefore at the meeting point

$$C(m_0, N) = D_1(m_0, N) = D_2(m_0, N)$$

which yields

$$N = \sqrt{\left(\frac{k^2}{2\pi} + \frac{1}{4}\right)} - \frac{1}{2} - \frac{1}{\sigma^2} \qquad (7.17)$$

$$\mu_N = m_0,$$

for the meeting point. All the risks at the sample size $N$ can now be calculated.

Now for any sample size $n$, one further observation $x$ leads to a posterior mean $\mu_{n+1}(\mu_n, x)$

$$\mu_{n+1}(\mu_n, x) = \{(n\sigma^2+1)\mu_n + x\sigma^2\}/\{(n+1)\,\sigma^2+1\}.$$

Therefore the continuation risk at $n$ can be written

$$C(\mu_n, n) = 1 + \int_{-\infty}^{\infty} \left\{ \int_{-\infty}^{\infty} \frac{1}{\sqrt{(2\pi)}} \exp\left\{-\frac{(x-m\ )^2}{2}\right\} . \frac{1}{\sqrt{(2\pi)}\,\sigma_n} \times \right.$$

$$\left. \exp\left\{-\frac{(m-\mu_n)^2}{2\sigma_n^2}\right\} \mathrm{d}m \right\} . R\{\mu_{n+1}(\mu_n, x), n+1\}\,\mathrm{d}x.$$

† This expression is the integral over $x$ of
(probability of the next obs. being $x$ integrated over $m$) ×
(Risk at sample point reached by increment $x$).
We could also write down

$$\int_{-\infty}^{\infty} \{\text{Prob of the next obs. being } x|m\} \times \{\text{Risk at the sample point reached } |m\}\,\mathrm{d}x$$

and then integrate over the posterior distribution for $m$ at $N$. Both formulations lead to the same result.

Since all the risks at sample size $N$ are known, the continuation risks at $(N-1)$ can be calculated, and all points at the sample size $(N-1)$ can be classified. Hence recursively, the decision boundaries can be worked back from the meeting point. This procedure is simple for an electronic computer, but may involve much time if $N$ is large.

The method just described for obtaining optimum sequential decision boundaries is quite general, so that there should be little difficulty in principle, in calculating optimum boundaries for any precisely defined decision theory model. However, there are considerable numerical difficulties in the procedure, arising from the build up of errors. Optimum boundaries for the binomial two decision problem of Ex. 7.8 have been given by Lindley and Barnett (1965), and these authors also give some consideration to the normal distribution case of Ex. 7.9.

## 7.5. Three Terminal Decisions

There is no difficulty in extending the above theory to problems involving three or more terminal decisions. Provided the decision boundaries for all neighbouring decisions meet, the whole $\xi$-space can be mapped by working back from the meeting-point boundaries, as before. Consider the following example.

*Ex. 7.10.* A two-sided test that binomial probability $\theta = \frac{1}{2}$ can be obtained as follows. Consider the prior distribution for $\theta$,

$$\begin{aligned} \Pr(\theta = \theta_1) &= a_1 \\ \Pr(\theta = \tfrac{1}{2}) &= a_2 \\ \Pr(\theta = 1-\theta_1) &= a_3 \end{aligned}$$

where

$$a_1+a_2+a_3 = 1, \quad \text{and} \quad \theta_1 > \tfrac{1}{2}.$$

Label the decisions that $\theta > \frac{1}{2}$, $\theta = \frac{1}{2}$, and $\theta < \frac{1}{2}$ as decisions 1, 2 and 3 respectively, and denote the loss of taking decision $i$ when $\theta$ is true as $W_i(\theta)$. (Let the cost of observations be the unit of costs.)

Ex. 7.10 as formulated will lead to open boundaries similar to those of Fig. 3.2. The boundaries can be closed by adding the condition that after $N_0$ observations, the cost of an observation is infinity. With this modification in the costs, the Bayesian solution will yield boundaries similar to those of Fig. 6.2a, except that the inner boundary will be larger.

Once the boundaries are obtained, they can be indexed and used in terms of probabilities of error if this is desired. This indicates one use of decision theory, namely that it is a very useful method for obtaining boundaries, which could then be indexed and used in other ways.

A similar formulation to Ex. 7.10 can be given for a two-sided test of a normal mean with known variance. In order to obtain a middle boundary for this problem, Armitage was forced to resort to an *ad hoc* rule, conjectures, etc., see § 6.5. It would be interesting to compare the properties of the boundaries Armitage obtained with those given by a decision theory formulation.

## 7.6. Empirical Bayes Solutions

Suppose a long sequence of batches of items is being inspected, and that the observation on each batch is a random variable $x_i$, which is normally distributed with unknown mean $m$ and unit variance. Let the prior distribution for $m$ be $N(\mu, \sigma^2)$, and let there be two terminal decisions, with loss functions as given in Ex. 7.9. Then by an approach similar to that used in Ex. 7.9, it can be shown that the optimum sample size for a single sample plan is a function $N(k, m_0, \mu, \sigma^2)$ of $k$, $m_0$, $\mu$ and $\sigma^2$. The boundary between the parts of the sample space in which either of the terminal decisions is preferred is the neutral line, and from Ex. 7.9 (see equation (7.15) and the following discussion), this is

$$m_0 = (N\sigma^2 \bar{x}_N + \mu)/(N\sigma^2 + 1),$$

where $\bar{x}_N$ is the sample mean.

Since there is a long sequence of batches being inspected, the parameters $\mu$ and $\sigma^2$ can be estimated from available data. For example, the grand mean of all observations (from all available batches), could be used as an estimator of $\mu$, and the between batches component of variance obtained from an analysis of variance to estimate $\sigma^2$. Thus although each batch is inspected by a single sample plan, both the neutral line and the sample size are functions (through $\mu$ and $\sigma^2$), of the results on previous batches, and in this sense the sampling plan is sequential. As the number of batches inspected increases, the estimates of $\mu$ and $\sigma^2$ become more precise, and provided the assumptions hold, the sampling plan tends asymptotically to one which is fully efficient.

This approach has been for a long time the way in which Bayes solutions were envisaged as being employed (see Barnard, 1954).

However, Robbins (1955, 1964) has recently put this empirical approach on a sound mathematical basis under the title 'Empirical Bayes procedures'. Robbins assumed only that the prior distribution belongs to a stated class of distributions, and he examines the conditions under which decision procedures which depend only on observed results on current and past batches have a risk asymptotically equal to the risk of the Bayes solution obtained when the prior distribution is known. If the class of prior distributions being used is severely restricted, as in the example discussed at the beginning of this section, no great problem is presented in the design of an empirical procedure. However, if the class of prior distributions is wide, there is at present no general method of constructing a suitable empirical procedure.

For an illustration of the difficulties we shall consider further the two decision problem based on a normal mean, which was discussed at the beginning of this section. The class of prior distributions for $m$ was taken to be $N(\mu, \sigma^2)$, which is rather a restrictive class in that it is unimodal. (In sampling inspection problems process curves often tend to have two or more modes.) Let us replace the prior distribution of $m$ by a density $g(m)$. After an observation $\bar{x}$ (on $n$ observations), the risks of taking the two terminal decisions are

$$k\int_{m_0}^{\infty}(m-m_0)\,g(m)\frac{\sqrt{n}}{\sqrt{(2\pi)}}\exp\left\{-\frac{n}{2}(\bar{x}-m)^2\right\}dm$$

and

$$k\int_{-\infty}^{m_0}(m_0-m)\,g(m)\frac{\sqrt{n}}{\sqrt{(2\pi)}}\exp\left\{-\frac{n}{2}(\bar{x}-m)^2\right\}dm$$

whence the neutral line is $(\bar{x}, n)$ satisfying

$$m_0\int_{-\infty}^{\infty}g(m)\frac{\sqrt{n}}{\sqrt{(2\pi)}}\exp\left\{-\frac{n}{2}(\bar{x}-m)^2\right\}dm$$

$$=\int_{-\infty}^{\infty}mg(m)\frac{\sqrt{n}}{\sqrt{(2\pi)}}\exp\left\{-\frac{n}{2}(\bar{x}-m)^2\right\}dm. \qquad (7.18)$$

The integral on the left of (7.18) is the probability density of $\bar{x}$, and can be estimated from back records by calculating the relative frequency of observations in a small interval near $\bar{x}$, and this estimate will be valid for any prior distribution. However, the integral on the right cannot be estimated without making some restrictive assumptions about $g(m)$. Robbins (1955) is able to work out the Poisson case, because the equivalent expression reduces to one similar to the expression on the left of (7.18); see the source paper for details.

A discussion of the choice of optimum $n$ introduces further difficulties, and the reader will readily check that an empirical estimate of this cannot be obtained when using the class of all prior distributions for $g(m)$. For practical purposes, therefore, it will nearly always be necessary to assume a parametric form for the prior distribution, and then estimate this from past data.

A further point to consider when specifying the class of prior distributions is the number of observations which will be required to obtain a satisfactory empirical Bayes procedure. For the normal example discussed above, if the prior distribution for $m$ is taken to be $N(\mu, \sigma^2)$, then good estimates of $\mu$ and $\sigma^2$ would be obtained by using a small number of observations, compared with the number required to determine the Bayes procedure on a wider class of prior distributions. Robbins's (1964) two-decision procedure for a Poisson mean requires a very large number of observations, whereas if it were possible to assume, say, a gamma prior distribution for the Poisson mean, much more efficient use of back records would result, or, alternatively, it would not be necessary to go back so far. If possible, we would prefer not to use the ancient history of a sequence in making a decision, because of the possibility of changes with time.

The last point leads to another possibility. Suppose we can assume that $x_i$ is normally distributed with unknown mean $m$ and unit variance, but that the prior distribution is only locally normal, $N(\mu, \sigma^2)$, there being slow changes in $\mu$ and $\sigma^2$ with time which can be ignored if not more than the previous $M$ batches are used to estimate $\mu$ and $\sigma^2$. Then $\mu$ could be estimated by a weighted moving average, and $\sigma^2$ estimated from a weighted average of squared differences between successive observations. In this way a single sampling plan could be constructed in which the sample size and neutral line vary in accordance with the current distribution of $m$.

## 7.7. The Minimax Criterion

Wald proposed that as an alternative to assuming a prior distribution, first the maximum of the risk $R(\theta|S)$ with respect to $\theta$ is found, and then the plan $S$ is chosen which minimizes this maximum,

$$\underset{S}{\text{Min}}\ \underset{\theta}{\text{Max}}\ R(\theta|S). \tag{7.19}$$

Wald showed that under certain very general conditions, the minimax solution is a Bayes solution with respect to the least favourable prior distribution, and this fact often provides a technique to obtain minimax solutions.

The minimax solution is criticized as being over-pessimistic. The values of $\theta$ for which $R(\theta|S)$ achieves a maximum may be known to occur only very rarely and in such a situation the minimax solution is governed completely by possibilities which are extremely unlikely to occur. More sensible results are sometimes obtained by a modification, called minimax regret. Regret is defined as the excess expected loss over the minimum possible if $\theta$ were known,

$$\{R(\theta|S) - \underset{S}{\text{Min}}\ R(\theta|S)\}.$$

The minimax principle is then applied to this quantity.

Often the minimax principle leads to mathematical difficulties, and in the few cases where the principle has been applied to obtain sequential procedures, extensive calculation has usually been necessary.

*Ex. 7.11* (Colton, 1963b). A number $N$ of patients have a certain disease, and they are to receive one of two treatments $A$ and $B$. An experiment is to be performed with some of the $N$ patients, to decide which treatment to use. We shall assume that the experiment is a paired comparisons design, and that the treatments are randomly assigned to the patients within each pair.

Let the responses to the two treatments be normally distributed random variables with known variance $\sigma^2$, but unknown means $\mu_A$ and $\mu_B$. The loss when a patient is given the inferior treatment is $C\theta$, where

$$\theta = \mu_A - \mu_B.$$

If $n$ pairs of patients are involved in the experiment the loss is

$$\text{Loss} = nC\theta \tag{7.20a}$$

if the experiment terminates with a decision for the superior treatment, and

$$\text{Loss} = C\theta\{n+(N-2n)\} \tag{7.20b}$$

otherwise. The cost of taking observations is considered to be of a different nature from the cost of administering the inferior treatment, and is therefore disregarded. Colton defends the use of a finite $N$ by saying that '... the decision between the treatments is not everlasting.'

When applying the minimax principle in the form (7.19) it is usual to limit the class $S$ of sampling plans under consideration. Colton considers the plan which continues the experiment until one of the following inequalities is broken,

$$-k\sigma^2 < \sum x_i < k\sigma^2, \tag{7.21}$$

where the $x_i$ are the differences of pairs of observations, and have a variance $2\sigma^2$. This sampling plan is an SPRT type, and some of its properties can be obtained by the theory of Chapter 2.

The SPRT for testing $H_0:\theta = -\phi$ against $H_1:\theta = \phi$ with probabilities of error $\alpha$, has the form

$$\frac{\sigma^2}{\phi}\log\left(\frac{\alpha}{1-\alpha}\right) < \sum x_i < \frac{\sigma^2}{\phi}\log\left(\frac{1-\alpha}{\alpha}\right),$$

so that we put

$$k = \left\{\log\left(\frac{1-\alpha}{\alpha}\right)\right\}\Big/\phi.$$

The OC-curve is obtained from Theorem 2.2, and is as follows.

$$P(\theta) = \frac{\left(\frac{1-\alpha}{\alpha}\right)^h - 1}{\left(\frac{1-\alpha}{\alpha}\right)^h - \left(\frac{\alpha}{1-\alpha}\right)^h}$$

where

$$h = -\theta/\phi.$$

This can be reduced to

$$P(\theta) = \frac{1}{e^{\theta k}+1}. \tag{7.22}$$

The ASN-function is given by Theorem 2.3, and this reduces to

$$E(n|\theta) = \frac{k\sigma^2}{\theta}\left(\frac{e^{\theta k}-1}{e^{\theta k}+1}\right). \tag{7.23}$$

From (7.20), the risk (expected loss) of this sampling plan is

$$E(\text{Loss}) = C\theta[E(n|\theta)+\{N-2E(n|\theta)\}\,\text{Pr(select inferior)}].$$

By substituting (7.22) and (7.23) we obtain

$$R(\theta|k) = C\theta\left\{\frac{k\sigma^2}{\theta}\left(\frac{e^{\theta k}-1}{e^{\theta k}+1}\right)^2 + N\frac{1}{(e^{\theta k}+1)}\right\}. \qquad (7.24)$$

To find the minimax solution, we find the maximum of (7.24) with respect to $\theta$, and then choose the value of $k$ which minimizes this maximum. This operation can be done by solving the equations

$$\frac{\partial}{\partial\theta}\{R(\theta|k)\} = 0$$

and

$$\frac{\partial}{\partial k}\{R(\theta|k)\} = 0.$$

Maurice (1959) has shown that this leads to

$$k = 0{\cdot}8262\sqrt{N}/\sigma\sqrt{2}$$
$$\theta = 2{\cdot}668(\sigma\sqrt{2})/\sqrt{N}.$$

By substituting these values into (7.23), the expected sample size of this plan is

$$E(n) = 0{\cdot}1241\,.N.$$

See also Colton (1963b) for the solution of this problem.

There are very few applications of minimax to sequential sampling. Maurice (1957) obtained the minimax Wald SPRT for testing hypotheses about an unknown normal mean ($\sigma^2$ known), under a slightly different model from the one used in Ex. 7.11 above. Breakwell (1956) considered minimax solutions for acceptance sampling, again using the SPRT. Anscombe (1958, 1961) applied the minimax regret criterion in obtaining rectifying inspection plans.

### 7.8 Discussion

The greatest value of the decision theory approach is most probably that it is a technique for obtaining decision boundaries. For some practical problems the formulation as such will be relevant, as in Ex. 1.3, while in others the boundaries could be indexed and used in terms of OC- and ASN-curves if this is desired.

One more criticism of the approach should be mentioned. Hill (1960, 1962, p. 42 and discussion, see also Whittle, 1954) puts forward the following idea. For inspection by consumers, the aim is not to achieve an economic optimum, but to encourage the producer to send the quality desired. Improved quality must cost more of course, so that, from the consumer's point of view, there is an optimum quality to buy. Hill argues that sampling inspection plans will influence the process curve (prior distribution) and sampling inspection tables should be designed to influence the producer's process curve in the desired direction. Hill's (1960) paper includes a simple illustrative example of how this idea might work out. From Hill's point of view the theory outlined in this chapter fails because it is not even designed to fulfil the most important criterion. No development of Hill's idea has been published yet.

## Problems 7

1. Suppose observations $x$ have a probability density function $\phi(x|\theta)$, and that the prior distribution for $\theta$ is $\xi(\theta|\alpha)$, which is closed under sampling. A sequential procedure is required to decide between two terminal decisions, $D_1$ and $D_2$, with losses $W_1(\theta)$, $W_2(\theta)$ respectively, where the cost of an observation (assumed constant) is taken as the unit of costs. Show that the neutral line is the locus of $\beta$ satisfying

$$\int \xi(\theta|\beta)\{W_2(\theta)-W_1(\theta)\}\,d\theta = 0.$$

Let $S(\beta,\theta)$ be the average sample size from any point $\beta$ in the $\xi$-space, when the optimum Bayes boundaries are specified, and when $\theta$ is true, and similarly, let $A(\beta,\theta)$ be the probability that sampling terminates with decision 1. Show that the decision 2 boundary is the locus of $\beta$ satisfying

$$\int [S(\beta,\theta)+A(\beta,\theta)\{W_1(\theta)-W_2(\theta)\}]\,\xi(\theta|\beta)\,d\theta = 0.$$

Obtain a similar equation for the decision 1 boundary, and show that if the decision 1 boundary meets the neutral line, then the decision 2 boundary does so at the same point. (See Wetherill, 1961.)

2. Determine an equation for the function $N(k, m_0, \mu, \sigma^2)$ of § 7.6, for

the optimum sample size of a single sample plan. Find an approximation for $N$.

3. Construct an empirical Bayes approach to the binomial acceptance sampling plan, Ex. 7.8. Outline an investigation into methods of estimation of the parameters $s$ and $t$, bearing the final discussion of § 7.6 in mind. How sensitive is the plan to sampling variations in $r$ and $s$?

4. Under what conditions will the two-sided test of a binomial probability outlined in principle in Ex. 7.10, lead to a closed sequential plan?

5. Horsnell (1957), has a different formulation of costs, which is along the following lines. Suppose batches of items are submitted for acceptance inspection, as a result of which they are either accepted or scrapped. A batch which is scrapped represents a loss of $k$ cost units, the unit being the cost of inspecting an item, which is supposed constant. Items inspected are classified as good or bad, and the prior distribution of $\theta$, the proportion of defectives in a batch, is $\xi(\theta)$. The average cost per batch accepted is

$$U = \{(\text{cost of inspection}) \text{ plus } (\text{cost of scrapping batches})(\text{proportion scrapped})\}/\{\text{proportion of batches accepted}\}.$$

Here we have

$$\text{Proportion of batches accepted} = \int_0^1 A(\theta)\,\xi(\theta)\,d\theta,$$

where $A(\theta)$ is the probability that a batch of quality $\theta$ is accepted. Horsnell suggests that $U$ is minimized subject to the restriction that when $\theta = \theta_1$, the probability of accepting a batch is less than $\beta$. Obtain an SPRT by this approach. Consider how you might apply this theory to the forest insect survey, Ex. 3.2.

What difficulties are encountered when attempting to obtain sequential plans by this approach, for a more general class of prior distributions than that leading to an SPRT?

6. Let $x_1$, $x_2$, ..., $x_n$ be independently and normally distributed

random variables with known variance $\sigma^2$ and unknown mean $\theta$. It is required to perform a two-sided test that $\theta = 0$, suitable for application to sequential medical trials (see § 6.2). Define suitable decision losses and a prior distribution for $\theta$, and obtain equations for the meeting points between the three decision regions for $\theta < 0$, $\theta = 0$ and $\theta > 0$. Show how to programme a computer to compute the Bayes boundaries, and to select a subset of boundaries for which the probability of saying $\theta = 0$ when this is true is at least $(1-\alpha)$.

Outline a set of investigations to compare the properties of your own solution to this problem with those in Chapters 3 and 6. Be careful to say what conditions you impose on two sets of boundaries before you regard them as equivalent for purposes of comparing average sample sizes, etc.

7. Colton (1963b) gives the following alternative formulation for Ex. 7.11. Each time a patient is given the superior treatment there is a gain of $C|\theta|$, where $C$ is some constant and $\theta$ the difference in means of the two treatments (see Ex. 7.11). Each time a patient is given the inferior treatment there is a loss of $C|\theta|$. If $2n$ patients are involved in the experiment, there is a gain of

$$\text{Gain} = \begin{cases} C(N-2n)|\theta| & \text{if the superior treatment is chosen,} \\ -C(N-2n)|\theta| & \text{if the inferior treatment is chosen.} \end{cases}$$

Use the sequential plan (7.21), and determine $k$ so that the minimum gain with respect to $\theta$ is maximized. (This approach is called 'Maximin Gain'.)

8. Colton (1963b) also gives a Bayesian approach to the determination of $k$ for the plan (7.21). Assume that the prior distribution for $\theta$ is normal with mean zero and variance $\sigma_0^2$. Use the Maximin Gain approach of the previous problem, and determine an optimum value of $k$.

9. Use the formulation of Ex. 7.7 with $f(n) = 0$, and derive the optimum binomial SPRT. Obtain the condition that both boundaries coincide with the neutral line. (See Vagholkar and Wetherill, 1960.)

## CHAPTER 8

# Sequential Estimation

### 8.1. Estimation versus Hypothesis Testing

A sequential procedure is defined by a stopping rule, which divides the sample space into a region in which further observations are taken, and one in which sampling is terminated. All the stopping rules discussed so far in this book have been obtained by formulating a problem as a test of simple hypotheses, or as a decision problem, to choose among a small number of courses of action. For many of the practical applications which have been mentioned the formulation as a problem of testing hypotheses is very artificial, and the parameters and risks of the formulation are very arbitrary; the hypothesis testing problem is merely used to obtain a stopping rule which appears reasonable. In many cases sequential estimation appears more appropriate, and offers many advantages over other formulations, even when a decision problem is evident. We briefly consider two examples.

Firstly consider sequential medical trials, discussed in Chapter 6. In any discussion of these in terms of hypothesis testing, it is assumed that a final decision to accept or reject a new treatment can be based simply on observed quantities $x_1$, $x_2$, ..., etc., on which a sequential test is based. In practice, situations are rarely so simple and the final decision depends on previous history of the treatments, side effects, cost and such factors, which are often difficult to weigh up. Armitage (1960, pp. 9–10) is careful to point this out, and he distinguishes between the conclusions made by the test and the policy decisions made by medical authorities. Even supposing the observed values $x_1$, $x_2$, ..., etc., were the only evidence relevant, a given set of results could be made to lead to different conclusions by varying the choice of $\alpha$, $\beta$ and the hypothesis values, and the non-statistician is asked to weigh up his decision problem not only before he sees any data, but also in terms of a choice of quantities he does not readily comprehend. It would be far more appropriate to require an investigator to present estimates of the difference in the effect of two treatments, with some estimate of the standard error, leaving the final decision to the medical authorities.

Hajnal (1960) also argues for an estimation approach to sequential medical trials. He says,

> 'Even if the use of a single criterion for the initial sequential analysis is accepted, an estimation procedure seems required rather than a decision procedure. Clinical trials are not normally followed by any once-and-for-all decision immediately on completion of the trial, and an estimate of the degree of superiority of a new treatment is usually an essential preliminary for considering the next step.'

Cox (1958, pp. 185–188) discusses Kilpatrick and Oldham's (1954) sequential medical trial, and raises the point mentioned above, that sequential estimation would be more appropriate for this kind of experiment. Two treatments for a bronchial disease were being compared, and effectiveness of treatments could be measured by the 'expiratory flow rate', denoted e.f.r. One treatment was a standard and the other a new one, and therefore the new treatment is only of interest if it is superior to the standard. Cox suggested that a suitable design would be one which estimated the $E$ (Difference in e.f.r.) with a standard error which depended on the value of the expected difference, in the manner shown in Fig. 8.1. An efficient estimation procedure satisfying this requirement will necessarily be sequential. Only after some information on the expected difference in the e.f.r. is available can the standard error required (and therefore the number of observations) be settled. (In the above discussion we have assumed that the variance of differences in the e.f.r. does not depend on the mean.)

In the discussion of Kilpatrick and Oldham's medical trial, Cox also refers to the severe limitations in the conclusions which can be drawn when stopping rules are obtained by the hypothesis testing approach. For example, all that can be concluded is that a null hypothesis is or is not rejected at a preassigned significance level, and the calculation of confidence intervals is difficult (see Armitage, 1958). The significance of the result is not affected by the number of observations taken to reach a decision, and it is not possible to assess the precision of comparisons with the results of other trials with the same treatments unless perhaps the other trials were done under the same conditions and with the same stopping rules as the trial under consideration. Since in most (if not all) medical trials other evidence will be obtained or will be available before reaching a final decision, this latter point is extremely important. When testing hypotheses by a fixed sample size plan, point and interval estimates can be incorporated into the conclusions, but the methods of estimation appropriate for sequential hypothesis-testing designs are not entirely clear; see § 8.3.

For a second example of an application suitable for sequential estimation, we refer to the forest insect survey, Ex. 3.2. Here the

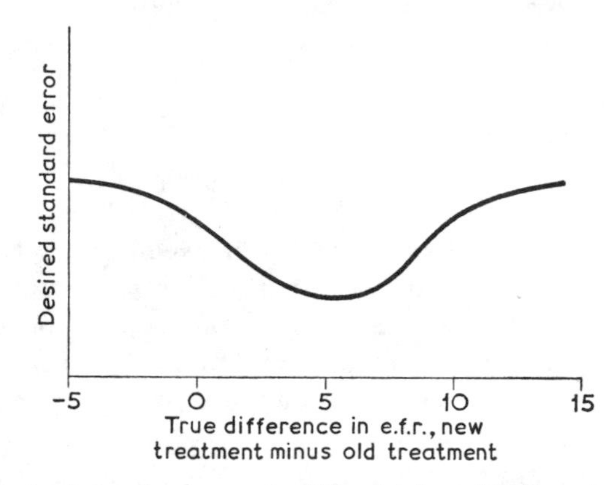


FIG. 8.1. Desired standard error for Kilpatrick and Oldham's experiment.

basic problem is where to spray, and how much, and Waters (1955) indicates that this decision may depend on the size of the area infested, whether parasites are present which control the insects, on economic considerations, etc. The decision will also presumably depend on conditions in neighbouring areas, previous history, and so on, and there might be several kinds of insect involved. To present the results of the survey merely as one of three simple statements that insect infestation is light, medium or heavy, seems to be an over-simplification, and Waters's claim that the SPRT method 'will make the best use of the competent observer's knowledge of an insect, the forest, and the basic dynamic relationships between them', is much too strong. It would be better to place upon the survey team the responsibility of providing estimates of the insect attack – the estimates to have a standard error which varies with the attack, and is large for low or high insect infestations but is smaller at intermediate levels of attack. Again, an efficient strategy must be sequential.

A sequential approach to estimation is appropriate only for certain situations, illustrated by the examples given above. We shall see later (§ 8.9) that the optimum sequential estimation procedure for some common estimation problems is a fixed sample size plan.

Two sets of considerations are involved in setting up a sequential estimation procedure.

(a) Design. The specification of the stopping rule, and comparisons between stopping rules.

(b) Inference. The rules by which an interval estimate, or point estimate and standard error, are to be made for each point where sampling may be terminated.

It is obvious that the considerations involved in (a) must depend on the way inferences are to be made. It is not agreed whether the method of making inferences should depend on the stopping rule, although most statisticians feel that it should. It is necessary to discuss some of the issues under argument before proceeding.

## 8.2. Estimation

There are two common modes of estimation, point estimation and interval estimation. By point estimation is meant a theory of deriving a single-valued function of a set of observations to represent an unknown parameter (or a function of unknown parameters), without explicitly stating the precision of the estimate. Normally, a point estimate without an associated standard error is of very little interest to the statistician, except in a decisions problem such as the following.

*Ex. 8.1.* Measurements of weight per unit length are taken on yarn produced by a certain machine. It is required to set the machine to produce yarn having a target value of weight per unit length, and the necessary adjustment to the setting must be calculated.

The adjustment in Ex. 8.1 is calculated from a point estimate of the difference between the actual and target mean weight per unit length. Decision estimates of this kind ought to be distinguished from the so-called theory of point estimation. If decision estimates are excluded, an interval estimate is in some ways an ideal form for an estimate, especially if this can be summarized as a point estimate and an associated standard error. This implies that interval estimation should be carried out first and then a convenient way of summarizing it found. In fact we frequently find that the reverse procedure is adopted – a point estimate is obtained, and then a rough point estimate of its standard error is obtained.

Both point and interval estimation can be based on either of two approaches, that through 'sampling distributions', or that through 'likelihood'. The sampling distribution approach bases all comparisons between estimators on the probability distribution of all possible results (over all sample sizes), and this of course depends on the stopping rule. For all likelihood approaches, including Bayesian, Bayesian decision theory and the (pure) likelihood theory of Barnard *et al.* (1962), inferences are conditional on the observed results and are therefore entirely independent of the stopping rule, once the observations have been obtained. (Barnard's theory does not involve point or interval estimates.) We shall outline the conventional sampling distribution approach first.

Suppose random variables $x_1, x_2, \ldots, x_n$ are observed, which are independent with a probability density $f(x, \theta)$, where $\theta$ is unknown. An estimate $\theta^*(x_1, \ldots, x_n)$ of $\theta$ is any function of the observed values which is calculated so as to be close to $\theta$. In order to formalize this requirement of 'closeness', we define a loss function $W(\theta^*, \theta)$, for the loss involved in estimating the true value $\theta$ by $\theta^*$. Some loss functions are

$$W_1(\theta^*, \theta) = k(\theta^* - \theta)^2$$

$$W_2(\theta^*, \theta) = k\,e^{-\alpha\theta^2}(\theta^* - \theta)^2$$

$$W_3(\theta^*, \theta) = k(\theta^* - \theta)^2/|\theta|$$

$$W_4(\theta^*, \theta) = k|\theta^* - \theta|^3$$

$$W_5(\theta^*, \theta) = k(1 - e^{-\alpha(\theta^* - \theta)^2})$$

$$W_6(\theta^*, \theta) = k(\theta^* - \theta)^2\,e^{-\alpha\theta^2} + k_2|\theta^* - \theta|$$

These loss functions have the property that for a given $\theta$, the loss increases as the difference between $\theta^*$ and $\theta$ increases, in either direction. However, $W_3$ is infinity when $\theta = 0$ and this would not usually be regarded as acceptable. Only $W_5$ or $W_6$ have what would normally be considered to be the right behaviour when $|\theta^* - \theta|$ is very large, that if the estimate is badly in error, a larger error in the same direction would not matter very much. Unfortunately, $W_4$ and

$W_6$ are very difficult to handle mathematically. The decision on a loss function for a particular application may be necessarily somewhat arbitrary.

Once a loss function is decided upon, we might consider trying to choose $\theta^*$ to minimize

$$\int_{\chi} W(\theta^*, \theta) \prod_{i} f(x_i|\theta)\,\mathrm{dx} \qquad (8.1)$$

for all $\theta$, where in sequential estimation, the integration is over the set $\chi$ of possible sets $(x_1, x_2, \ldots)$ which are terminal results of an experiment. No estimator $\theta^*$ exists having minimum risk uniformly in $\theta$, for if $\theta = d$, then $\theta^* = d$, independent of the observations, has minimum risk, but only when $\theta = d$. Therefore it is not satisfactory to allow the estimator $\theta^*$ to be any function, and one or more conditions will have to be imposed on it. We can then seek to find the $\theta^*$ minimizing (8.1) among this restricted class of estimators. Here another arbitrary step is involved, for there are several restrictions which can be used, and there is no general agreement that any particular one is preferred to all others. The most frequently quoted restrictions are unbiasedness, invariance, and consistency.

For sequential estimation, an unbiased estimate is one for which the expectation over $\chi$ is equal to the true value of the unknown parameter. Approximate unbiasedness is often important, but exact unbiasedness arises almost exclusively because it gives rise to the easiest mathematics. There may be estimates with slight bias which have a much smaller expected loss. Fisher (1956), p. 140 says

> 'A primary, and really very obvious, consideration is that if an unknown parameter $\theta$ is being estimated, any one-valued function of $\theta$ is necessarily being estimated by the same operation. The criteria used in the theory must, for this reason, be invariant for all such functional transformations of the parameters. This consideration would have eliminated such criteria as that the estimate should be "unbiased" ...'

One example of this is estimation of the variance $\sigma^2$ of a normal distribution with unknown mean. The standard procedure uses $s^2$ as a point estimate, where

$$s^2 = \frac{1}{n-1} \sum (x_i - \bar{x})^2,$$

and the factor $(n-1)$ is defended by showing that $E(s^2) = \sigma^2$. However, it is the standard deviation which is nearly always required, and yet, since

$$\{E(s)\}^2 + V(s) = E(s^2) = \sigma^2$$
$$E(s) < \sigma.$$

The restrictions upon estimators alternative to unbiasedness include invariance, consistency and median unbiasedness. There are general discussions on these points in a number of advanced statistical textbooks, such as Kendall and Stuart (1961), and Fraser (1957), and the discussions are readily adapted to apply to sequential estimation.

The theory of sequential point estimation differs very little in principle from fixed sample size estimation, but the details are more complicated to work out. An estimator $\theta^*$ is found which minimizes (8.1) subject to one or more restrictions of the type mentioned above, and once an estimator is decided upon, its sampling variance can be determined, and a sample estimate of this can be associated with each point estimate as a measure of error. Clearly, both the estimate and the standard error will depend critically on the stopping rule, that is, upon what we might have done had sampling terminated at a point other than that observed. This is the argument discussed in § 6.7, which leads many statisticians to reject methods of estimation based on sampling distributions, and favour methods based on likelihood. The argument cannot be given much weight in fixed sample size plans, for there seems no reason to consider other stopping rules, but in sequential sampling the argument has considerable force. Two examples are worked out in § 8.3 and § 8.4, illustrating the various points under discussion here.

Barnard (see Barnard *et al.*, 1962, and references) has frequently advocated the direct study of the likelihood as a procedure for estimation. If $N$ observations with a sample mean $\bar{x}$ are known to be from a normal distribution with variance $\sigma^2$, then the likelihood function of the unknown mean $\theta$ is proportional to

$$\exp\left\{-\frac{N(\bar{x}-\theta)^2}{2\sigma^2}\right\}, \tag{8.2}$$

which is a function of $\theta$, conditional on $\bar{x}$ and $N$ and is independent of the stopping rule. Thus for estimation by Bayesian and Bayesian decision theory approaches, as well as for the method suggested by

Barnard *et al.* (1962), based on likelihood alone, no special methods of estimation are needed for sequential sampling. Methods of choosing suitable stopping rules are required, of course, but these in no way affect estimation once the observations are obtained.

Suppose that for the likelihood (8.2), we assume a prior distribution $d\theta$ over the whole real line, then the posterior distribution of $\theta$ is

$$\frac{\sqrt{N}}{\sqrt{(2\pi)}\,\sigma}\exp\left\{-\frac{(\theta-\bar{x})^2 N}{2\sigma^2}\right\}, \tag{8.3}$$

which is symmetrical about $\bar{x}$ with a variance $\sigma^2/N$. In this sense we might refer to $\sigma/\sqrt{N}$ as the spread of the likelihood. Now it so happens that for a fixed sample size plan, the sampling distribution of $\bar{x}$ has the same form as (8.3), so that point estimates and standard errors will be the same for likelihood and sampling distribution approaches. When this happens we might argue that the difference in philosophical approach to estimation alters our interpretation of results but not the sampling procedures. For the normal distribution example given here it is easy to see that of all stopping rules which could be drawn through $(\bar{x}, N)$, only the fixed sample size plan yields a sampling distribution of $\bar{x}$ of the same form as (8.3).

A detailed discussion of confidence intervals will not be given here, and some theory is incorporated into the discussion of the examples given in the next two sections. The most important property for our present discussion is that confidence intervals depend on the stopping rule, whereas interval estimates based on a likelihood approach do not.

## 8.3. Inverse Binomial Sampling

Suppose independent observations $x_i$ are taken on a Bernoulli process in which $x_i = 1$ with probability $\theta$, and $x_i = 0$ with probability $(1-\theta)$, and

$$\sum_{i=1}^{n} x_i = r,$$

say. If a fixed number $n$ of observations is taken, the standard error of the estimate $(r/n)$ of $\theta$ is $\sqrt{\{\theta(1-\theta)/n\}}$, and this is approximately $\sqrt{\theta}/\sqrt{n}$ if $\theta$ is small. If the estimate of $\theta$ is required to have a standard error approximately proportional to $\theta$, a method of sampling is required which takes more observations for smaller values of $\theta$, than for $\theta$ nearer $\frac{1}{2}$. Haldane (1945), Tweedie (1945) and others have discussed a procedure known as inverse binomial sampling, which roughly achieves this standard error requirement; the procedure is simply to continue taking observations until a fixed number $c$ of positive results has been observed. A number of authors have discussed estimation from this procedure, taking different points of view, and inverse sampling is an important example for illustrating the different views taken to estimation in general, and sequential estimation in particular.

Let $n$ denote the number of observations taken until the $c$th positive result occurs, then the probability distribution (that is, the sampling distribution) of $n$ is

$$\Pr\{n = m|\theta, c\} = \binom{m-1}{c-1}\theta^c(1-\theta)^{m-c} \tag{8.4}$$

where $m = c, c+1, c+2, \ldots$ The maximum likelihood estimator of $\theta$, $c/m$, is biased, see Haldane (1945). No simple closed form exists for $E(c/m)$ for general $c$, but it is easily shown that

$$\theta < E(c/m) < \{c/(c-1)\}\theta, \tag{8.5}$$

see problem P8.6. Thus the bias tends to zero for large values of $c$, but may be considerable at small values of $c$. For $c = 2$, the expectation can be evaluated exactly and we have

$$E\left(\frac{2}{m}\right) = 2\left\{\frac{\theta}{(1-\theta)} + \frac{\theta^2}{(1-\theta)^2}\log\theta\right\}. \tag{8.6}$$

Some calculated values are shown in Table 8.1. The bias can be very considerable. (In fact, the reader will readily check that $m/c$ is an unbiased estimate of $1/p$.)

TABLE 8.1. $E(2/m)$ for inverse binomial sampling, $c = 2$

| $\theta$ | 0·01 | 0·10 | 0·20 | 0·30 | 0·40 | 0·50 |
|---|---|---|---|---|---|---|
| $E(2/m)$ | 0·0192 | 0·1654 | 0·2988 | 0·4148 | 0·5188 | 0·6138 |
| % bias | 92 | 65 | 50 | 39 | 30 | 23 |

From (8.4) we have

$$E\{(c-1)/(m-1)\} = \theta, \tag{8.7}$$

and the estimator $(c-1)/(m-1)$ has minimum variance among all unbiased estimators of $\theta$.

Haldane (1945), see also Kendall and Stuart (1961, p. 594), showed that

$$E\{(c-1)^2/(m-1)^2\} = \theta^2\left\{1+\frac{1-\theta}{c}+\frac{2(1-\theta)^2}{c(c+1)}+\ldots\right\} \quad (8.8)$$

and that an unbiased estimate of the variance of $(c-1)/(m-1)$ is

$$\text{Est } V\{(c-1)/(m-1)\} = \frac{(c-1)(m-c)}{(m-1)^2(m-2)}. \quad (8.9)$$

These properties of the estimator $(c-1)/(m-1)$ are true only for the distribution (8.4), which results from inverse binomial sampling. Consider the case $c = 2$ and suppose that in a particular example sampling is terminated at $m = 12$, then if the stopping rule is changed in the region of $m = 100$, the distribution (8.4) no longer holds, and the unbiased estimator with minimum variance is not $(c-1)/(m-1)$, nor is the variance of the estimator (8.9). (In this case only small changes are made in the estimator and its variance. However, if $c = 2$ is changed to $c = 3$ for $m \leqslant 11$, and $c = 2$ for $m \geqslant 12$, then a very different unbiased estimator results for the same observed point ($c = 2, m = 12$).) Therefore when estimates are formed by taking expectations over the sample space, they depend not only on what *was* observed, but on what might conceivably have been observed, but wasn't.

Barnard *et al.* (1962) quote inverse sampling as an example illustrating how wrong (according to them!) it is to employ methods of estimation based on sampling distribution properties. The likelihood, they say, contains *all* the information, and the stopping rule is irrelevant to methods of estimation and inference. The likelihood is the probability of observing the results obtained, as a function of the unknown parameter $\theta$. For binomial sampling this is proportional to

$$\theta^c(1-\theta)^{m-c}$$

irrespective of whether $m$ or $c$ were fixed in advance. A maximum likelihood estimator of $\theta$ is $(c/m)$ *independently* of the stopping rule, and from Table 8.1 we see that this estimator has an expectation over the sample space with up to 100% positive bias. There are a number of ways in which the 'likelihood' statistician can argue round this, but first we face another issue.

We can obtain a measure of spread of the likelihood as follows. Let $\theta$ have a uniform prior distribution on $(0,1)$, then when sampling is terminated at $(c,m)$ the posterior distribution for $\theta$ is a

$$\beta(c+1, m-c+1)$$

distribution, with a mode at $c/m$, and expectation and variance

$$E(\theta) = \frac{(c+1)}{(m+2)}$$

and

$$V(\theta) = \frac{(c+1)(m-c+1)}{(m+2)^2(m+3)}. \tag{8.10}$$

If $c = 2$, $m = 3$, formula (8.9) gives 1/4 while (8.10) gives 1/25. The ratio of (8.9) to (8.10) does not tend to unity for large $m$, but to $(c-1)/(c+1)$. This situation contrasts markedly with the normal distribution example of § 8.2, in which the sampling distribution and the posterior distribution had the same form (8.3). For inverse binomial sampling, the approaches to estimation via sampling distributions and via likelihood lead in general to different estimates and variances. However, we might observe that on the basis of the likelihood function, there is very little basis for choosing between the $c/m$ and $(c-1)/(m-1)$ estimators. Consider the set of estimates

$$\frac{c+z}{m+z} = 1+\left(\frac{c-m}{m+z}\right), \tag{8.11}$$

where $z \geqslant 1-c$, and $m \geqslant c$. As $z$ tends to infinity, these estimates tend to unity independently of $c$ and $m$. If $z = -c$, all estimates are zero unless $m = c$, when they are indeterminate. The ratio of the likelihoods for estimate (8.11) to estimate $(c/m)$ is

$$R(z,m,c) = \frac{\left(1+\frac{z}{c}\right)^c}{\left(1+\frac{z}{m}\right)^m}. \tag{8.12}$$

Now the function

$$Y(z|\omega) = \left(1+\frac{\omega}{z}\right)^z$$

for $z > 0$, and $\omega \geqslant 1-z$, is an increasing function of $z$. Thus the minimum value with respect to $m$, of the likelihood ratio (8.12) is attained as $m \to \infty$, yielding

$$\operatorname*{Min}_{m} R(z, m, c) = \left(1+\frac{z}{c}\right)^{c} e^{-z}, \tag{8.13}$$

which holds for all $z \geqslant 1-c$. Values of (8.13) are given in Table 8.2.

If m $= c$, all estimates (8.11) are equal, so the maximum value of (8.12) for $m$ is unity for every $z$. Apart from this special case, the likelihood ratios rapidly approach the limit (8.13) as $m$ increases.

This argument shows that there is always a region near $(c/m)$ and $(c-1)/(m-1)$ in which almost any point may quite likely be the true value of $\theta$. The differences between estimates obtained from the same data by different approaches reveals the extremely arbitrary nature of 'point estimation'. The large differences in the variances (8.9) and

TABLE 8.2. Minimum values of the ratio of the likelihoods for estimates $(c+z)/(m+z)$ to the likelihood for the estimate $(c/m)$

| | $c$ | | | |
|---|---|---|---|---|
| $z$ | 2 | 3 | 4 | 5 |
| 5 | 0·0825 | 0·1278 | 0·1727 | 0·2156 |
| 4 | 0·1649 | 0·2327 | 0·2931 | 0·3462 |
| 3 | 0·3112 | 0·3983 | 0·4670 | 0·5221 |
| 2 | 0·5412 | 0·6264 | 0·6850 | 0·7277 |
| 1 | 0·8278 | 0·8721 | 0·8982 | 0·9155 |
| 0 | 1 | 1 | 1 | 1 |
| −1 | 0·6796 | 0·8054 | 0·8601 | 0·8907 |
| −2 | – | 0·2737 | 0·4618 | 0·5746 |
| −3 | – | – | 0·0783 | 0·2057 |
| −4 | – | – | – | 0·0175 |

(8.10) is more troublesome and the combined effect of consistent differences in point estimates and the associated variances is clearly not a negligible feature.

There are difficulties in deciding for either method of estimation in connection with inverse binomial sampling. Estimates through likelihood may have undesirable sampling properties, while any estimates through sampling distributions are influenced by what was not observed. An alternative answer is that to ask for a point estimate in a general sense (without using an infinite number of observations), is an absurd question. Point estimates will be required for some purpose, and if suitable loss functions can be defined, a Bayesian or minimax approach could be adopted both to the determination of the estimates and of the stopping rule. This approach will in effect attribute most of the biases to the stopping rule, rather than to the method of estimation. A Bayesian approach, of course, results in estimates depending on the likelihood and not on the stopping rule, and the prior distribution performs the role of a weight function, to combine the losses for different values of $\theta$. The only non-Bayesian approach which appears suitable is through Anscombe's (1953) asymptotic theory, which shows how to construct boundaries so that the customary (fixed sample size) inferences are valid to a given degree of approximation.

We conclude this section with some comments on confidence intervals for the inverse binomial sampling stopping rule. A lower $\alpha\%$ confidence bound is the value of $\theta$ satisfying

$$\Pr\{n \leqslant m|\theta, c\} = \alpha$$

or

$$\theta^c + c\theta^c(1-\theta) + \frac{c(c+1)}{1.2}\theta^c(1-\theta)^2 + \ldots + \binom{m-1}{c-1}\theta^c(1-\theta)^{m-c} = \alpha.$$

However, from (1.1) this probability statement is equivalent to

$$\sum_{r=c}^{m} \binom{m}{r} \theta^r(1-\theta)^{m-r} = \alpha$$

which is

$$\Pr\{r \geqslant c|\theta, m\} = \alpha,$$

where sampling is binomial. Therefore lower $\alpha\%$ confidence bounds for direct and inverse sampling are identical. Upper $\alpha\%$ confidence bounds would be obtained by considering $\Pr\{n \geqslant m|\theta, c\}$ and $\Pr\{r \leqslant c|\theta, m\}$. However, it follows from the above discussion that

$$1 - \Pr\{n \leqslant m|\theta, c\} = 1 - \Pr\{r \geqslant c|\theta, m\}$$

hence

$$\Pr\{n > m|\theta, c\} = \Pr\{r < c|\theta, m\}.$$

Now

$$\Pr\{n \geqslant m|\theta, c\} = \Pr\{n > m|\theta, c\} + \binom{m-1}{c-1}\theta^c(1-\theta)^{m-c}$$

and

$$\Pr\{r \leqslant c|\theta, m\} = \Pr\{r < c|\theta, m\} + \binom{m}{c}\theta^c(1-\theta)^{m-c}.$$

Therefore unless $c = m$, $\Pr\{n \geqslant m|\theta, c\}$ is not equal to $\Pr\{r \leqslant c|\theta, m\}$, and the upper confidence limits given by direct and inverse binomial sampling will be different. (See Barnard *et al.*, 1962, and Morris, 1963.) In this case the difference in confidence intervals is due to inclusion or exclusion of the point $r = c$, $n = m$. However, confidence intervals are not in general identical for different stopping rules, and this can be seen by considering a stopping rule slightly different from inverse sampling.

Consider the stopping rule which samples until we satisfy

$$\sum_{i=1}^{n} x_i = \begin{cases} 2 \ (n \leqslant 3) \\ 3 \ (n \geqslant 4) \end{cases} \tag{8.14}$$

where $x_i$ is zero or unity, as defined at the beginning of this section. Armitage (1958) described how confidence limits can be obtained for some types of boundaries defined on Bernoulli trials. The method is basically that given above for inverse sampling, but first, the possible stopping points have to be put in an ordered sequence depending on the estimate of $\theta$ (this ordering depends on the estimate). The conditions for confidence intervals to exist are that the distribution function over the boundary points is a monotonic function of $\theta$. The confidence intervals for stopping rule (8.14) are left as a problem to the reader (P8.8). It is easily seen that confidence intervals for the point $(\sum x_i = 3, n = 5)$ are not identical to those for binomial sampling.

## 8.4. Estimation of a Normal Mean

Before proceeding with the theory we briefly consider another problem. Suppose observations $x_i$, for $i = 1, 2, \ldots$, are independently and normally distributed with a known variance $\sigma^2$ and an unknown mean $\theta$. Suppose an estimate of $\theta$ is required which is to have high precision if $|\theta|$ is small, then it may be reasonable to use the stopping rule defined as follows: take observations until the inequality

$$\left|\sum_{i=1}^{n} x_i\right| > k \tag{8.15}$$

is satisfied, where $k$ is a positive constant. This stopping rule is called inverse normal sampling, and approximations to some of its properties can be derived by using the diffusion theory approximation described in the Appendix. We replace the sample size variable $n$ by a continuous variable $t$, so that at termination the sample mean is either $k/t$ or $-k/t$. In order to examine the sampling distribution of the mean at termination, we must derive the distribution of $1/t$ for a diffusion process with two absorbing barriers.

Let the random variable $x$ be subject to independent normal increments with a mean $\theta \Delta t$ and variance $\sigma^2 \Delta t$ in a time interval $\Delta t$. Let there be two absorbing barriers, at $x = a$ and $x = -a$, and let the process start from $x = 0$ at a time $t = 0$. The probability density of the process being at $x$ at a time $t$ can be obtained by the method of images, see Bartlett (1962, Chapter 3), and this is

$$p(x,t) = \frac{1}{\sigma\sqrt{(2\pi t)}} \sum_{-\infty}^{\infty} \left[ \exp\left\{ \frac{\theta x_n'}{\sigma^2} - \frac{(x - x_n' - \theta t)^2}{2\sigma^2 t} \right\} - \exp\left\{ \frac{\theta x_n''}{\sigma^2} - \frac{(x - x_n'' - \theta t)^2}{2\sigma^2 t} \right\} \right], \qquad (8.16)$$

for $-a \leqslant x \leqslant a$, where the images are

$$\left.\begin{aligned} x_n' &= 4na \\ x_n'' &= 2a - 4na \end{aligned}\right\}$$

for $n = 0, \pm 1, \pm 2, \ldots$ By integrating (8.16) with respect to $x$ in the interval $(-a, a)$ and then differentiating with respect to $t$, the probability density of the time $t$ at termination is obtained in the form

$$\frac{a}{\sigma\sqrt{(2\pi)}\, t^{3/2}} \sum_{-\infty}^{\infty} (4n-1) \exp\left\{ -\frac{(4n-1)^2 a^2}{2\sigma^2 t} \right\}, \qquad (8.17)$$

for $\theta = 0$. The probability density of $1/t$ can be obtained by multiplying (8.17) by $t^2$, and this results in a density which is a sum of gamma densities of the type

$$\frac{\sqrt{k} z^{-1/2} \mathrm{e}^{-kz}}{\sqrt{\pi}},$$

where $z = 1/t$, which is infinite at $z = 0$. Therefore when $\theta = 0$, the probability density of $t^{-1}$ is infinite at $t^{-1} = 0$.

By following through the same algebra for $\theta \neq 0$, the probability density of $t^{-1}$ is obtained as a sum of terms such as

$$\frac{1}{\sqrt{(2\pi)}} \exp\left\{-\frac{(4na-a+\theta t)^2}{2\sigma^2 t}\right\}.\frac{(4na-a+\theta t)}{\sigma t^{3/2}}$$

and the density of $t^{-1}$ is zero at $t^{-1} = 0$. These arguments show that the sampling distribution of $\bar{x}$ from the stopping rule (8.15) is as shown in Fig. 8.2a for $\theta = 0$ and Fig. 8.2b for $\theta > 0$. The sampling distribution of $\bar{x}$ is in any case a compound of the separate sampling distributions for each boundary.

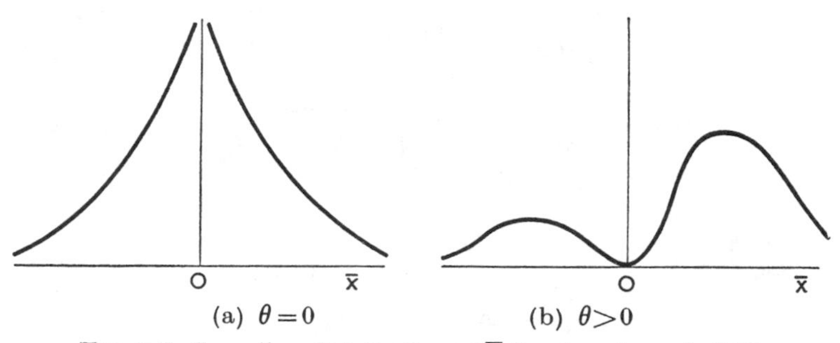


FIG. 8.2. Sampling distribution of $\bar{x}$ for stopping rule 8.15.

There are three sampling distributions which could be considered as a basis for inferences about $\theta$:

(a) The completely unconditional sampling distribution, shown in Fig. 8.2.

(b) The sampling distribution conditional on the boundary reached.

(c) The sampling distribution conditional on both the boundary reached and the sample size. This is approximately a truncated normal distribution.

We should usually like inferences to be as highly conditional as possible, subject to there being no loss of information. Most of the information in the observations lies in quantities $k/n$ or $-k/n$, depending on which boundary a sample terminated. Since sampling distribution (c) is conditional on $k$ (or $-k$) and $n$, most of the information in the data is eliminated by this approach, and it is plainly wrong. (For further objections to sampling distribution (c) see problem P8.1.)

It is more difficult to decide between sampling distributions (a) and (b) and a strong case could be made for using the boundary to determine the sign of an estimate, and then using sampling distribution (b) to make a point estimate of $|\theta|$. The bimodal nature of sampling distribution (a) for $\theta > 0$ causes some difficulty. In other respects a discussion of point estimation from stopping rule (8.15) follows closely upon that for inverse binomial sampling, given in § 8.3 above, and the same criticism applies, that the inference made will depend to some extent on what is not observed. The force of this criticism can be seen from the following argument.

Suppose $k$ is $O(\sigma)$, and that the lower boundary of the stopping rule can be ignored for some range of $\theta > 0$. The distribution of $t$ for one boundary can be obtained by differentiating equation (A.6) of the Appendix, which yields

$$\frac{k}{\sigma\sqrt{(2\pi)}\, t^{3/2}} \exp\left\{-\frac{\theta^2 t}{2\sigma^2}+\frac{k\theta}{\sigma^2}-\frac{k^2}{2\sigma^2 t}\right\}.$$

This is the inverse Gaussian distribution studied by Tweedie (1957), who showed that

$$E\left(\frac{1}{t}\right) = \frac{\theta}{k}+\frac{\sigma^2}{k^2},$$

whence

$$E\left(\frac{k}{t}\right) = \theta+\frac{\sigma^2}{k},$$

and the bias in an estimate $k/t$ of $\theta$ is $O(\sigma)$. If the boundary is rotated clockwise about the observed point $(\bar{x}, n)$ until it becomes vertical, then (ignoring the lower boundary), this bias diminishes to zero. If the boundary is rotated anticlockwise about $(\bar{x}, n)$, the bias in the sample mean is rapidly increased.

Confidence intervals for $\theta$ from a Neyman–Pearson standpoint would be derived as follows. The joint probability density of $\bar{x}$ and $n$ must be of the form

$$\frac{C(\bar{x}, n)\sqrt{n}}{\sqrt{(2\pi)}\,\sigma} \exp\left\{-\frac{n(\bar{x}-\theta)^2}{2\sigma^2}\right\}$$

where $C(\bar{x}, n)$ is a combinatorial term resulting from integrations over all paths leading to $(\bar{x}, n)$ which satisfy the stopping rule. The constant

$C(\bar{x},n)$ does not involve $\theta$, and a uniformly most powerful test of $H_0:\theta=\theta_0$ against $H_1:\theta>\theta_0$ is of the form

$$\frac{\exp\left\{-\frac{n}{2\sigma^2}(\bar{x}-\theta)^2\right\}}{\exp\left\{-\frac{n}{2\sigma^2}(\bar{x}-\theta_0)^2\right\}} \gtrless \mathrm{d}(\bar{x},n)$$

where $d$ must be chosen to give the test a required size $\alpha$. In order to choose $d$, the constant $C(\bar{x},n)$ has to be determined, so that $d$ is a function of $\bar{x}$ and $n$ (and the stopping rule). Suppose now we mark off on the $(\bar{x},n)$ plane the area consistent with $\theta=\theta_0$ at the significance level $\alpha$, and then do this for all values of $\theta_0$. The confidence interval for $\theta$ for an observed $(\bar{x},n)$, is the set of all $\theta_0$ for which the observed result $(\bar{x},n)$ is not significant at the given level $\alpha$. These confidence intervals have not been evaluated, but it is clear that they will differ considerably from the confidence intervals obtained from a fixed sample size, with the same observations, because of the form of the sampling distribution of $\bar{x}$. If the upper boundary is rotated about an observed point $(\bar{x},n)$, the confidence intervals will be displaced. It would be of interest to see actual numerical results of confidence intervals obtained on this type of stopping rule.

To the likelihood statistician, the stopping rule (8.15) is irrelevant to inferênce. If the sample size at termination is $m$, and the sum of the observations is $X_m$, the likelihood function of $\theta$ is proportional to

$$\exp\left\{-\frac{(X_m-m\theta)^2}{2\sigma^2 m}\right\}$$

whence an estimate of $\theta$ is $(X_m/m)$, the sample mean. If the prior distribution for $\theta$ is taken to be uniform on $(-\infty,\infty)$, then 95% of the posterior distribution of $\theta$ lies between

$$\{X_m/m \pm 1{\cdot}96\sigma/\sqrt{m}\}.$$

These results hold regardless of the stopping rule, and their sampling distribution properties can be altered by changing it. In particular the sample mean $X_m/m$ is a biased estimate of $\theta$ for stopping rule (8.15). The likelihood statisticians' reply about these sampling distribution properties is in terms of the ratio of likelihoods, parallel to the discussion in § 8.3 for inverse binomial sampling.

## 8.5. Discussion of § 8.2 to § 8.4

The examples discussed in § 8.3 and § 8.4 have been mentioned repeatedly in papers on the foundations of statistical inference, see Barnard *et al.* (1962), and Cornfield's discussion of Birnbaum (1962). In the statistics of fixed sample size procedures such discussions are very largely of academic interest only, but in sequential estimation it is not possible to take this attitude. Apart from the asymptotic theory described below, there is no common ground where the same procedures would be used regardless of different interpretations.

Much of the dispute arises out of an unnecessary confusion of the related – but distinct – problems of designing a stopping rule and making an inference. The design of stopping rules involves a study of the sampling distribution properties of likelihoods, or of point estimates, and the probability that interval estimates will contain the true values, etc. A method of estimation can also be based on expectations over the sample, but I consider that this is not reasonable, at least in a sequential sampling plan. On the other hand, there are 'peculiar' stopping rules, such as those given in § 8.3 and § 8.4, for which many statisticians would doubt the unqualified use of the likelihood principle.

In a decision theory approach to sequential estimation of a normal mean (variance $\sigma^2$ known), I conjecture that the only conditions under which the boundaries of § 8.4 are admissible, in the decision theory sense, are with a two point prior distribution for the mean $\theta$, when the cost of observations is constant. Unless we are in the very unusual situation where $\theta$ is known to be either of two values $\theta_1$ or $\theta_2$, a continuous prior distribution must be assumed, and the boundaries will meet, so that the plan of § 8.4 is not optimum. In fact, a decision theory approach to sequential estimation problems will lead to procedures consistent with the likelihood principle for which the customary inferences (in fixed sample sizes) are approximately correct. Before describing the decision theory approach, we shall survey some standard results.

## 8.6. Some Theoretical Results

If we consider that possible bias in estimators from sequential procedures is important, then it would be useful to have a method of constructing unbiased estimators. Girshick, Mosteller and Savage

(1946) gave a method of obtaining unbiased estimates of $\theta$ (and of some other functions such as $\theta^2$, $\theta(1-\theta)$, etc.), from any sequential procedure on Bernoulli trials (see § 8.3). Some general theory leading up to a result by Cox (1952b) gives a method of constructing approximately unbiased estimates of an unknown mean from a more general class of sequential procedures. Neither of the methods claims to yield estimates with any optimal variance property.

Suppose we have any sequential procedure defined on Bernoulli trials, that is we have a boundary or set of boundaries in the space $(n,r)$, where $r$ is the number of positive results in $n$ trials. We shall assume that the boundaries are such that, whatever $\theta$, the probability of reaching a boundary point is unity. Let $(s,t)$ be any boundary point, or any other point which can be reached from the origin by a path not touching a boundary. Then if $(n,r)$ is a boundary point, denote by $a(n,r|s,t)$, the number of paths from $(s,t)$ to $(n,r)$ which do not touch a boundary. We shall prove that an unbiased estimate of $\theta^t(1-\theta)^{s-t}$ is

$$E\left\{\frac{a(n,r|s,t)}{a(n,r|0,0)}\right\} = \theta^t(1-\theta)^{s-t} \tag{8.18}$$

where expectations are taken over all boundary points $(n,r)$. Now the expectation is

$$\begin{aligned} E\left\{\frac{a(n,r|s,t)}{a(n,r|0,0)}\right\} &= \sum \frac{a(n,r|s,t)}{a(n,r|0,0)} \cdot a(n,r|0,0)\, \theta^r(1-\theta)^{n-r} \\ &= \sum a(n,r|s,t)\, \theta^r(1-\theta)^{n-r}, \end{aligned} \tag{8.19}$$

where the summation extends over all boundary points $(n,r)$. There are no permissible paths from $(s,t)$ to $(n,r)$ if $n < s$ or $r < t$. For all other $(n,r)$, the probability of a path from $(s,t)$ to $(n,r)$ is

$$a(n,r|s,t)\, \theta^{r-t}(1-\theta)^{n-r-s+t}.$$

Now since the probability of reaching a boundary point from the origin – and thus from any reachable point $(s,t)$ – is assumed to be unity, we have

$$\sum a(n,r|s,t)\, \theta^{r-t}(1-\theta)^{n-r-s+t} = 1.$$

But from (8.19)

$$\begin{aligned} \sum a(n,r|s,t)\, \theta^r(1-\theta)^{n-r} &= \theta^t(1-\theta)^{s-t} \sum a(n,r|s,t)\, \theta^{r-t}(1-\theta)^{n-r-s+t} \\ &= \theta^t(1-\theta)^{s-t}, \end{aligned}$$

which proves (8.18).

*Example 8.2.* For inverse sampling, $c \geqslant 2$, the number of paths to $(n, c)$ is $\binom{n-1}{c-1}$, while the number of paths from $(1,1)$ to $(n, c)$ is $\binom{n-2}{c-2}$. The unbiased estimate of $\theta$ is therefore

$$\frac{\binom{n-2}{c-2}}{\binom{n-1}{c-1}} = \frac{c-1}{n-1}.$$

Similarly, an unbiased estimate of $\theta^2$ is

$$\frac{(c-1)(c-2)}{(n-1)(n-2)},$$

provided $c \geqslant 3$. This leads to the unbiased estimate of $V(\theta)$ given in equation (8.9).

## 8.7. Asymptotic Boundary Theory

Anscombe (1953) considers the problem of estimating the mean of a normal distribution by a confidence interval of specified width, when the variance is unknown. Suppose $x_i$, $i = 1, 2, \ldots, n$, are independent random variables from a normal distribution with expectation $\theta$ and variance $\sigma^2$, then write

$$u_i = \frac{1}{i(i+1)}\left\{ix_{i+1} - \sum_{j=1}^{i} x_j\right\}^2, \tag{8.20}$$

for $i = 1, 2, \ldots, (n-1)$, so that the $u_i$ are independently distributed as $\sigma^2 \chi^2$ on one degree of freedom. After $n$ observations, the width of the fixed sample size $(1-\alpha)\%$ confidence intervals is $2l$, where

$$l^2 = \frac{t_\alpha^2}{n(n-1)} \sum_1^{n-1} u_i$$

and where $t_\alpha$ is the percentage point of the $t$-distribution. If we can assume that fixed sample size formulae hold, then an approximate sequential procedure is to stop sampling as soon as

$$\sum_1^{n-1} u_i < (nl/\lambda_\alpha)^2, \tag{8.21}$$

where $\lambda_\alpha$ is the percentage point of the normal distribution.

Anscombe's second-order theory leads to modifying the stopping rule (8.21) to

$$\sum_{i=1}^{n-1} u_i < \left(\frac{nl}{\lambda_\alpha}\right)^2 \left(1 - \frac{2{\cdot}676}{n} - \frac{\lambda_\alpha^2}{2n}\right) \tag{8.22}$$

for $n \geqslant 4$.

The method Anscombe used was to assume a knowledge of the skewness of the distribution of the $u_i$, and then write down the approximate Edgeworth type expansion for the distribution of

$$U_n = \sum_1^n u_i,$$

in terms of the skewness of the $u_i$ and the unknown variance of the normal variables $x_i$. The boundary is assumed to be of the form $U_n = f(n)$, where

$$f(n) = cn^\beta \left(1 + \frac{b}{n} + \ldots\right)$$

and $\beta$ is different from unity. Under certain assumptions, the sample size distribution $\Pr\{n < N\}$ can now be evaluated as the unrestricted probability that $U_N$ has crossed the boundary, plus the probability that $U_j$ crossed the boundary for some $j < N$, but has crossed back again. Both of these separate probabilities are obtained by approximate asymptotic arguments.

From the asymptotic sample size distribution, the first four moments can be obtained, and these can be used to obtain an approximation to the distribution of

$$\bar{x} = \frac{1}{n} \sum_1^n x_i,$$

unconditionally on the sample size at which an experiment is terminated. This is done by expanding the distribution of $\bar{x}$ given $n$ about a value $v$ which is close to $E(n|\theta, \sigma^2)$ and then using the known moments of $n$. This distribution of $\bar{x}$ involves the parameters of the boundary, and the confidence statement is correct to the order of approximation used if a certain condition is fulfilled. The stopping rule (8.22) satisfies this condition. Other applications of the theory can be given.

We may remark that for those who like inferences to be highly conditional, or for those who base inferences only on the likelihood, the results of § 8.6, § 8.7 and § 8.8 below do not convey much of value. Perhaps the most important result is that, in large sample sizes, sequential estimation differs little from fixed sample size estimation.

It seems appropriate to quote Anscombe's own comments on his results. He says (1953, p. 18), 'I cannot remember any occasion, however, when the methods given above ... would have seemed fit and appropriate to recommend.'

Since 1953 the use of electronic computers has become widespread and Tocher (1964) has recommended Anscombe's method for certain simulation and empirical sampling trials. Sequential estimation procedures are likely to become increasingly important in connection with applications of computers.

## 8.8. Lower Bound to the Variance of Sequential Estimates

The fixed sample size result on the lower bound to the variance of an estimate was extended to sequential sampling by Wolfowitz (1947). We give a sketch of the proof, and for details and regularity conditions, etc., see the source paper.

Suppose the distribution of independent random variables $x_i$, $i = 1, 2, \ldots,$ is $f(x, \theta)$, so that

$$\int f(x, \theta)\, \mathrm{d}x = 1.$$

Under certain conditions we may differentiate to obtain

$$\int \frac{\partial f(x, \theta)}{\partial \theta} \mathrm{d}x = 0 = \int \frac{\partial \log f(x, \theta)}{\partial \theta} f(x, \theta)\, \mathrm{d}x = E\left\{\frac{\partial \log f(x, \theta)}{\partial \theta}\right\}.$$

Now put

$$Y_n = \sum_{i=1}^{n} \frac{\partial \log f(x_i, \theta)}{\partial \theta},$$

then it follows from Wolfowitz (1947) that

$$E(Y_n) = E(n|\theta) E\left(\frac{\partial \log f(x, \theta)}{\partial \theta}\right) = 0, \tag{8.23}$$

and

$$V(Y_n) = E(n)\, E\left\{\frac{\partial \log f(x, \theta)}{\partial \theta}\right\}^2. \tag{8.24}$$

Let an estimator of $\theta$ be $\theta^*(x_1,\ldots,x_n)$, then

$$E\{\theta^*(x_1,\ldots,x_n)\} = \theta+b(\theta) \tag{8.25}$$

where the expectation again extends over all sets $(x_1,\ldots,x_n)$ and $n$ which are possible as terminal results of the sequential procedure. Denote the set of possible results $\{x_1,\ldots,x_n\}$ for a sample size $n$ by $\chi_n$, then (8.25) is

$$\theta+b(\theta) = \sum_{r=1}^{\infty}\int_{\chi_r} \theta^*(x_1,\ldots,x_r)f(x_1|\theta)\ldots f(x_r|\theta)\,\mathrm{d}x_1\ldots\mathrm{d}x_r,$$

and by differentiation we have

$$1+b'(\theta) = E\left\{\theta^*(x_1,\ldots,x_n)\,.\sum_{i=1}^{n}\frac{\partial\log f(x_i,\theta)}{\partial\theta}\right\} = E\{\theta^*\,Y_n\}. \tag{8.26}$$

By applying a well-known inequality to this we obtain

$$V(\theta^*)\,V(Y_n) \geqslant \{C(\theta^*,Y_n)\}^2 = \{E(\theta^*\,Y_n)\}^2,$$

and by using (8.24) we have the final result

$$V(\theta^*) \geqslant \frac{\{1+b'(\theta)\}^2}{E(n|\theta)\,E\{\partial\log f(x,\theta)/\partial\theta\}^2}. \tag{8.27}$$

For inference conditional on the sample size, the usual fixed sample size version of the bound applies.

This result can be expressed as an inequality on the expected number of observations. Therefore, if for any given estimation problem there is a fixed sample size plan which achieves the desired variance, and this sample size is known, then a sequential plan will not lead to a saving of observations (on average) over the fixed sample size plan. The result (8.27) also shows that if the standard error and sample size desired depend on the value being estimated, as in the examples discussed in § 8.1, then we might hope at best to find a sequential plan which is on average as efficient as if we had known what sample size to take.

### 8.9. Bayesian Decision Theory Approach to Sequential Estimation

The problem of sequential estimation was mentioned by Wald (1947), and one special problem was solved by him later (see P8.11), in a decision theory setting.

Suppose we have to obtain a point estimate of the probability $\theta$ of Bernoulli trials, then the discussion in § 8.2 shows that there are two questions to investigate, the determination of the estimator, and the determination of a stopping rule. In a decision theory approach, the estimator used at a given terminal point is independent of the stopping rule, and dependent only on the posterior distribution.

Consider inverse binomial sampling (§ 8.3), then an estimator of $\theta$ is required, which is some function $\theta^*(c,m)$, of the sufficient statistics $c$ and $m$, and the standard approach to estimation at present requires $\theta^*$ to have some suitable sampling distribution property. For example, we may want to consider estimators $\theta^*(c,m)$ which have small expected squared error and we consider the function

$$R_1(\theta^*, c, \theta) = \sum_{m=c}^{\infty} \binom{m-1}{c-1} \theta^c (1-\theta)^{m-c} \{\theta^*(c,m)-\theta\}^2. \quad (8.28)$$

No function $\theta^*(c,m)$ exists which has a smaller $R_1(\theta^*,c,\theta)$ than any other $\theta^*$ uniformly in $\theta$, and we look for estimators $\theta^*$ which give a smaller risk $R_1$ among a restricted class of estimators; see § 8.2.

The decision theory approach might be as follows. Let the prior density of $\theta$ be $f(\theta)$ for $0 \leqslant \theta \leqslant 1$, and let the loss incurred in taking an estimate $\theta^*(c,m)$ when $\theta$ is true be $k(\theta^*-\theta)^2$, where the cost of an observation is the unit of costs. The risk of inverse sampling is (provided the order of integration and summation can be reversed),

$$R_2(\theta^*, c) = \sum \binom{m-1}{c-1} \left[ \int_0^1 \theta^c (1-\theta)^{m-c} f(\theta) \{m + k(\theta^*-\theta)^2\} \, d\theta \right].$$

We seek to minimize this risk with respect to choice of both the estimator $\theta^*$ and the stopping rule. (For inverse sampling, determination of the stopping rule reduces to choosing a suitable value for $c$.) The estimator can be determined independently of the stopping rule since $R_2(\theta^*,c)$ is minimized for choice of $\theta^*$ if

$$\int_0^1 \theta^c (1-\theta)^{m-c} f(\theta) (\theta^*-\theta)^2 \, d\theta$$

is minimized for every $m \geqslant c$. Suppose $f(\theta)$ has a beta distribution with parameters $\alpha$ and $\beta$, then this integral is proportional to

$$\theta^{*2} - 2\theta^* \frac{\alpha+c}{\alpha+\beta+m} + \frac{(\alpha+c)(\alpha+c+1)}{(\alpha+\beta+m)(\alpha+\beta+m+1)}$$

which is minimized if

$$\theta^* = \frac{\alpha+c}{\alpha+\beta+m}. \tag{8.29}$$

This estimator is obtained conditionally on results $(c, m)$, and so does not depend on the stopping rule. Having determined $\theta^*$, we can now proceed to determine the optimum value of $c$, and this clearly depends on the constant $k$, which performs the role of balancing estimation losses against the cost of sampling. The determination of the optimum $c$ is left as a problem to the reader, see P8.7.

It is easy to see that it will be generally true that the Bayes estimator can be determined independently of the choice of the stopping rule. The cost of sampling cannot depend on $\theta^*$, so that only estimation losses are relevant to the choice of $\theta^*$. Suppose we have distributions closed under sampling, then with the notation of Chapter 7, a stopping rule is the specification of subsets $\chi_j$, for $j = 1, 2, \ldots,$ of the sample space in which sampling is terminated at sample size $j$. Let the loss of estimating the true value $\theta$ by $\theta^*$ be $W(\theta^*, \theta)$, then the overall risk is

$$R_3(\theta^*, \alpha) = \sum_j \int_{\chi_j} \int W(\theta^*, \theta)\, \phi(x_1|\theta) \ldots \phi(x_j|\theta)\, \xi(\theta|\alpha)\, \mathrm{d}\theta\, \mathrm{d}x$$

where the first integration ranges over all subsets $\{x_1, \ldots, x_j\}$ leading to termination of sampling at a sample size $j$. $R_3$ will be minimized for choice of $\theta^*$ if the integral

$$U(\theta^*, \beta) = \int W(\theta^*, \theta)\, \xi(\theta|\beta)\, \mathrm{d}\theta$$

is minimized for choice of $\theta^*$, for all reachable points $\beta$ of the $\xi$-space.

Thus an estimator can be determined by decision theoretical considerations, without regard to the stopping rule to be used. Such estimators depend only on the likelihood through the posterior distribution, and on the loss function $W(\theta^*, \theta)$. If, however, the loss function is multiplied by a constant the estimator will be unaltered.

Having determined an estimator, there remains the problem of determining an optimum stopping rule. For example, we can consider a series of risks such as $R_2$, for estimation of binomial $\theta$, employing different stopping rules (but the same estimator), and we can ask which rule gives the minimum overall risk. This problem is now similar to those discussed in § 7.4, and similar methods can be used to obtain the optimum boundaries. However, there are an infinite number of possible decisions here, corresponding to all possible estimates of $\theta$, compared with only two decisions in § 7.4, and neither the neutral boundary nor the meeting point boundary can be defined as in Chapter 7.

A sequential estimation procedure is a division of the $\xi$-space into two regions, a continuation region, and a termination region, plus an estimation rule for the termination region. As soon as the termination region is reached sampling is stopped and the point estimate $\theta^*(\beta)$ is made, where $\beta$ is the point in $\xi$-space. Let $\xi_n$ be the subset of the $\xi$-space corresponding to all points reachable at a sample size $n$ from a given starting point. Suppose the cost of sampling is constant and the unit of costs, and write the continuation risk at $\beta_n$ $C\{\theta^*(\beta_n), \beta_n\}$, where $\beta_n$ is in $\xi_n$, then we have a similar set of equations to those obtained in Chapter 7. The basic equation is

$$C(\beta_n) = 1+\int \left\{ \int_{-\infty}^{\infty} \xi(\theta|\beta_n)\, \phi(x'|\theta)\, \mathrm{d}\theta \right\} R(\beta'_{n+1})\, \mathrm{d}x, \qquad (8.30)$$

where

$$R(\beta) = \mathrm{Min}[D\{\theta^*(\beta), \beta\}, C(\beta)] \qquad (8.31)$$

and

$$\xi(\theta|\beta'_{n+1}) = \frac{\xi(\theta|\beta_n)\, \phi(x'|\theta)}{\int_{-\infty}^{\infty} \xi(\theta|\beta_n)\phi(x'|\theta)\, \mathrm{d}\theta}$$

and where $D\{\theta^*(\beta), \beta\}$ is the termination risk at $\beta$. (The estimator $\theta^*(\beta)$ is written into the decision risk to emphasize that the termination risk depends on the estimation procedure used, as well as on $\beta$. There is no reason why $\theta^*(\beta)$ should be chosen to be the function exactly minimizing the posterior risk, see the discussion of Ex. 8.5 below.)

Frequently there will be a maximum reachable sample size, and if this is determined, the optimum boundaries can be found by working backwards using (8.30) and (8.31). Let us define the incremental gain at $\beta_n$ as the reduction in estimation risk obtained by sampling just one more observation,

$$\Delta(\beta_n) = D\{\theta^*(\beta_n), \beta_n\} - \int \left\{ \int_{-\infty}^{\infty} \xi(\theta|\beta_n)\, \phi(x'|\theta)\, \mathrm{d}\theta \right\} D\{\theta^*(\beta'_{n+1}), \beta'_{n+1}\}\, \mathrm{d}x$$

then the meeting point can be defined as the $\beta_N$ for which

$$1 = \underset{\beta_n \epsilon \xi_n}{\text{Max}} \ \Delta(\beta_n). \qquad (8.32)$$

where unity is the cost of one observation. (There is an obvious generalization when the cost of observations is a function of $n$.) The neutral boundary could be defined as the locus of maximum incremental gain among subsets $\xi_n$, and if the problem has symmetry this reduces to the set of points $\beta_j$ for which

$$D(\theta^*(\beta_j), \beta_j\} = \underset{\beta_j \epsilon \xi_j}{\text{Max}} \ D\{\theta^*(\beta), \beta\}.$$

One special case which is of interest is when the termination risk $D\{\theta^*(\beta), \beta\}$ and the incremental gain $\Delta(\beta)$ are functions of $n$ only and so are constant within each subset $\xi_j$. When this arises the optimum sequential stopping rule is a fixed sample size plan. Ex. 8.3 below illustrates this situation.

Another special case is when the function $\Delta$ is independent of $n$, or alternatively if the cost of observations is not constant, $\Delta$ is a function of $n$ which cancels out in (8.32). Wald (1950) gave an illustration of this, for estimation of the unknown mean of a rectangular distribution of unit length; see P8.11. By assuming the cost of observations constant, the incremental gain was found to be independent of $n$. This situation results in a stopping rule independent of $n$. See problem P8.2 for another example of this special case.

*Ex. 8.3.* Suppose $x_i$, $i = 1, 2, \ldots$, are independently normally distributed with known variance $\sigma^2$, and unknown mean $\theta$. Let the prior distribution for $\theta$ be $\mathrm{d}\theta$, $-\infty < \theta < \infty$, and the loss function

$$W(\theta^*, \theta) = k(\theta^* - \theta)^2$$

where the cost of an observation is the unit of costs.

The posterior distributions for $\theta$, given an observed sample mean $\bar{x}$, is normal with mean $\bar{x}$ and variance $\sigma^2/n$. The optimum estimator will be the $\theta^*$ which minimizes

$$k \int_{-\infty}^{\infty} (\theta^* - \theta)^2 \frac{\sqrt{n}}{\sqrt{[(2\pi)\,\sigma]}} \exp\left\{-\frac{n}{2\sigma^2}(\theta - \bar{x})^2\right\} \mathrm{d}\theta$$

whence $\theta^* = \bar{x}$, and the minimized posterior risk is $k\sigma^2/n$, which is independent of $\bar{x}$.

The optimum stopping rule has a meeting point which is found by equating the posterior risk at $n$, to one plus the risk at a sample size $(n+1)$, based on the posterior distribution of $\theta$ at $n$. This is

$$\frac{k\sigma^2}{N} = 1 + \frac{k\sigma^2}{N+1},$$

and the $[N]$ given by this equation defines the meeting point.

Since the risk given $\bar{x}$ is independent of $\bar{x}$, the optimum scheme is a single sample plan in which $N$ observations are taken.

This last result corresponds to one obtained by Stein for confidence intervals, in which the optimum procedure is also a fixed sample size one. It appears that for certain types of commonly occurring estimation problems, sequential procedures do not lead to any increase in efficiency, and therefore would usually be undesirable. However, we mentioned in § 8.1 that there is a class of problems in which an efficient procedure will inevitably be sequential. Consider the following problem.

*Ex. 8.4.* Suppose we have the set-up given in Ex. 8.3, except that the loss function is

$$W(\theta^*, \theta) = k\,\mathrm{e}^{-\frac{\alpha\theta^2}{2\sigma^2}}(\theta^* - \theta)^2. \tag{8.33}$$

With this loss function, the loss corresponding to a given $(\theta^* - \theta)^2 = d^2$, say, diminishes as $\theta$ moves away from the origin. This loss function might be appropriate for sequential medical trials.

It could be argued that the loss function should be modified slightly to weight differently for $|\theta^*| > \theta$ than for $|\theta^*| < \theta$, however, the function (8.33) can be used as an approximation in a preliminary analysis.

The posterior risk at $(\bar{x}, n)$ is

$$R(\theta^*|\bar{x}, n) = k \int_{-\infty}^{\infty} \exp\left\{-\frac{\alpha\theta^2}{2\sigma^2}\right\} . (\theta^* - \theta)^2 \frac{\sqrt{n}\exp\left\{-\frac{n}{2\sigma^2}(\theta - \bar{x})^2\right\}}{\sqrt{(2\pi)}\sigma}\,\mathrm{d}\theta$$

$$= \frac{k\sqrt{n}}{\sqrt{(n+\alpha)}}\exp\left\{-\frac{\alpha n\bar{x}^2}{2(n+\alpha)\sigma^2}\right\} \int (\theta^* - \theta)^2 \frac{\sqrt{(n+\alpha)}}{\sqrt{(2\pi)}\,\sigma} \times$$

$$\exp\left\{-\frac{(n+\alpha)}{2\sigma^2}\left(\theta - \frac{n\bar{x}}{n+\alpha}\right)^2\right\}\mathrm{d}\theta. \tag{8.34}$$

It is clear that (8.34) is minimized if we choose $\theta^* = \bar{x}/(1+\alpha/n)$. We might expect the value of $\alpha$ used to be small, say less than $\sigma^2$, in which case the minimum risk estimate is very close to $\bar{x}$. It is doubtful if in practice any estimate other than $\bar{x}$, would be used. The minimized values of the posterior risks are

$$R(\theta^* = \bar{x}|\bar{x}, n) = \frac{k\sqrt{n}}{\sqrt{(n+\alpha)}}\left\{\frac{\alpha^2\bar{x}^2}{(n+\alpha)^2}+\frac{\sigma^2}{(n+\alpha)}\right\}\exp\left\{-\frac{\alpha n\bar{x}^2}{2(n+\alpha)\sigma^2}\right\} \tag{8.35}$$

and

$$R\left(\theta^* = \frac{n\bar{x}}{n+\alpha}\middle|\bar{x}, n\right) = \frac{k\sqrt{n}}{\sqrt{(n+\alpha)}}\left\{\frac{\sigma^2}{(n+\alpha)}\right\}\exp\left\{-\frac{\alpha n\bar{x}^2}{2(n+\alpha)\sigma^2}\right\} \tag{8.36}$$

so that at $\bar{x} = 0$, the risks are identical, and equal to $k\sqrt{n}\,\sigma^2/(n+\alpha)^{3/2}$. These risks are a maximum at $\bar{x} = 0$, and it is clear that an optimum boundary will have its largest value of $n$ at $\bar{x} = 0$.

The posterior probability density from $(\bar{x} = 0, n)$, that the next observation is $x$, is

$$\frac{\sqrt{n}}{\sqrt{(2\pi)}\,.\,\sqrt{(n+1)}\,\sigma}\exp\left\{-\frac{nx^2}{2(n+1)\sigma^2}\right\}, \tag{8.37}$$

and by using (8.35) and (8.37), the continuation risk at $(\bar{x} = 0, n)$ for the estimate $\theta^* = \bar{x}$ is seen to be equal to

$$1+\frac{k\sqrt{n}\sigma^2}{\sqrt{(n+\alpha)}\,(n+1+\alpha)}\left\{1+\frac{\alpha^2}{(n+1)^2\,(n+\alpha)}\right\} \tag{8.38}$$

if $(\bar{x} = 0, n)$ is a meeting point. By equating expressions (8.35) at $\bar{x} = 0$ and (8.38), we obtain the sample size $N$ corresponding to the meeting point.

From the meeting point, we can work backwards, equating continuation risks with the risk of stopping and making an estimate, and determine the stopping rule optimum for $\theta^* = \bar{x}$. (Notice that the boundaries corresponding to $\theta^* = \bar{x}/(1+\alpha/n)$ will be slightly different. These can be obtained in a similar way.)

Before using a stopping rule determined as indicated above, it would be desirable to know some sampling properties of the procedure, such as sampling variations in the shape of the likelihood at termination, the distribution of $(\bar{x}, n)$ at termination, particularly for $\theta = 0$, and also some characteristics of the sample size distribution would be of interest. Further, since the parameters $k$ and $\alpha$ will usually be known only roughly, it would be better to compare a few alternative sets of boundaries obtained by varying $k$ and $\alpha$, before deciding on a set to use. With an electronic computer, these investigations are fairly trivial.

An interesting application of the above theory to obtain stopping rules is given by Freeman (1972). The problem considered is the sequential estimation of the size of a population in a capture-recapture experiment. The decision theoretic formulation enables him to arrange a balance between the cost of observations and losses due to inaccuracy of estimation. (See also El-Sayyad and Freeman (1973).)

## 8.10. Discussion

When a likelihood approach to estimation is adopted, the stopping rule can be ignored when choosing an estimator, but there remains a problem of choosing 'reasonable' stopping rules. Decision theory provides an ideal framework to choose stopping rules, and the calculation of the boundaries is straightforward when an electronic computer is available. The effect of variations in the prior distribution and costs, and the sampling properties of the boundaries should be studied before any plan is employed.

If some care is not taken in the choice of the stopping rule, we might easily choose a stopping rule with a high expected loss, due either to high sample sizes or to boundaries yielding a poor sampling distribution of the likelihood function. An alternative method of avoiding 'unsuitable' boundaries is by Anscombe's (1953) asymptotic theory, but the decision theory approach is of more general applicability.

In many cases Amster's (1963) modified Bayes rule will be satisfactory. In this method, the continuation risk is computed at every point as if a fixed number of further observations are to be taken. (For example, take one more observation and then make an estimate.) Another development which might be useful in a few applications is Robbins (1964) empirical Bayes procedures (see § 7.6). There appears to be a need for a great deal of exploratory work in the field covered in this chapter.

## 8.11 Sequential Interval Estimation

So far in this chapter we have been considering point estimation, and we now briefly discuss some of the work on sequential interval estimation.

The problem of obtaining a confidence interval of specified width and confidence coefficient for the mean of a normal distribution of known variance was considered by Wald and Stein (1947). These authors showed that no sequential procedure existed for this problem which had an average size less than that of the classical fixed sample size procedure. This result seems intuitively reasonable, but it also seems clear that sequential methods should be able to help in cases such as Kilpatrick and Oldham's medical trial, discussed in § 8.1, where the width of a confidence interval depends on its location. Accordingly Paulson (1969) reformulated the problem of obtaining sequential confidence intervals for a normal mean in the following way.

Observations are taken on independent, normally distributed random variables of known variance $\sigma^2$. A confidence interval ($L_n$, $U_n$) is required such that

(i) $U_n - L_n \leqslant W$
(ii) $U_n - L_n \leqslant W + \lambda L_n \quad$ if $L_n > 0$
(iii) $U_n - L_n \leqslant W - \lambda U_n \quad$ if $U_n < 0$

where $n$ is the number of observations and $\lambda$ an arbitrary constant. This set of requirements leads to confidence intervals of width less than $W$ if zero is included, but of increasing width as the distance between zero and the confidence interval increases.

The method depends on the normal diffusion process result given in the appendix. From (A.6), the probability of crossing a boundary at $A$ before $T$ is

$$2\{1 - \Phi(A/\sigma\sqrt{T})\}$$

Therefore for random variables $x_1, x_2, \ldots$, having a normal distribution $N(\mu, \sigma^2)$,

$$\Pr\left\{\sum_{1}^{n}(x_i - \mu) > A \text{ for at least one } n, 1 \leqslant n \leqslant T\right\} \leqslant 2\{1 - \Phi(A/\sigma\sqrt{T})\}$$

If we now write $Z(p)$ such that

$$\Phi\{Z(p)\} = 1 - p$$

and choose $T_1$ to be the smallest integer greater than

$$4\sigma^2 Z^2(\alpha/4)/W^2$$

and put $A_1 = WT_1/2$, we can see that

$$\Pr\left\{-A_1 \leqslant \sum_1^n (x_1 - \mu) \leqslant A_1 \text{ for all } n,\ 1 \leqslant n \leqslant T\right\}$$

$$\geqslant 1 - 4\{1 - \Phi(A_1/\sigma\sqrt{T})\} = 1 - \alpha$$

Therefore, if we define

$$L_n = \max\ (1 \leqslant i \leqslant n)\ \{\bar{x} - A_1/r\}$$

and

$$U_n = \min_i\ (1 \leqslant i \leqslant n)\ \{\bar{x} + A_1/r\}$$

then

$$\Pr\{L_n \leqslant \mu \leqslant U_n \text{ for all } n,\ 1 \leqslant n \leqslant T_1\} \geqslant 1 - \alpha$$

Finally, we notice that the following stopping rule is a satisfactory solution to the problem: stop sampling as soon as either

(i) $U_n - L_n \leqslant W$

or (ii) $L_n > 0$ and $U_n - L_n \leqslant W + \lambda L_n$

or (iii) $U_n < 0$ and $U_n - L_n \leqslant W - \lambda L_n$

or (iv) $n = T_1$.

The source paper, by Paulson (1969) gives an application of this to obtaining sequential confidence intervals for the difference between two normal means, and an extension using the $t$-distribution to deal with the case when $\sigma^2$ is unknown. Empirical sampling trials demonstrate that a substantial saving in sample size is possible. However, there are parameter configurations for which the average sample size is larger than that of the equivalent fixed sample size procedure.

Sequential confidence intervals have been derived in certain other cases. O'Brien (1973) has obtained a procedure for sequential interval estimation of the shape parameter of the gamma distribution, and Cooke (1971) gave a sequential confidence interval for the parameter of the uniform density on $(0, \theta)$. For some of the more theoretical aspects of sequential estimation see Knight (1965), Linnik and Romanovsky (1972), Bickel and Yahav (1969) and references.

**Problems 8**

1. Discuss estimation for inverse normal sampling, § 8.4, by sampling distribution (c), and consider the following points.
(i) Obtain the distributions of $(x_1-k)$ and $(x_1+x_2-k)$, conditional on the process stopping at the upper boundary at sample sizes 1 and 2 respectively.
(ii) What effect does variation in $k$ have on the maximum likelihood estimators of $\theta$ from the distributions obtained in (i)?
(iii) Suppose the process stopped at sample size 1 with $x_1 = X > k$, and at sample size 2 with $(x_1+x_2) = 2X > k$, and obtain estimates of $\theta$ from each result. How different are these estimates?

2. Let random variables $x_i$ have a Poisson distribution, with $\Pr(x=r) = e^{-m} m^r/r!$, for $r = 0, 1, 2, \ldots$, and let the prior density of $m$ be

$$\frac{\alpha^{\beta+1}}{\Gamma(\beta+1)} m^{\beta} e^{-\alpha m}, \qquad m \geqslant 0.$$

Let the cost of estimating $m$ by $\hat{m}$ be $k(\hat{m}-m)^2/m$, where the cost of taking an observation (assumed constant) is taken as the unit of costs. Show that the optimum decision boundaries are a fixed sample size procedure.

Under what conditions on the costs will the decision boundaries have the form $\sum x_i = K$?

3. What is the loss in efficiency in the problem described in Ex. 8.4, in using $\bar{x}$ instead of $\bar{x}/(1+\alpha/n)$? How do you measure this loss in efficiency?

4. Formulate a decision theoretical sequential estimation approach suitable for the forest insect survey, Ex. 3.2. Detail what properties of the stopping rule you would investigate, including sampling distribution properties, and say how you would compare your plan with alternatives. (What alternatives?) When actually using one of your plans, would you choose a plan entirely by decision theoretic considerations, or would you insist on some probability-of-error type of restriction?

5. Formulate a decision theoretical sequential estimation approach to sequential medical trials. (Consult the papers by Anscombe,

Armitage, and Colton, in 1963 *J.A.S.A.*) Outline a method of comparing your plans with those of Colton (1963b) and Anscombe (1963).

6. For inverse binomial sampling, discussed in § 8.3, show that

$$E\left(\frac{c}{m}\right) < E\left(\frac{c}{m-1}\right) = \frac{c}{c-1}.\theta.$$

Also show that

$$\frac{m-1}{m} \geqslant \frac{c-1}{c}$$

for $m \geqslant c$, and hence show that

$$E\left(\frac{c}{m}\right) > \theta.$$

7. Decision theory calculations were started in § 8.9 to obtain a Bayes estimator of $\theta$ and an optimum value of $c$ for inverse binomial sampling (see equations (8.28), (8.29), and the surrounding text). Show that if a uniform prior distribution is taken for $\theta$ in the range $(0, 1)$, there is no value of $c$ giving finite expected loss.

8. Suppose you have used the stopping rule (8.14), and reached the point $\sum x_i = 3$, $n = 5$, where $\Pr(x_i = 1) = \theta$, and $\Pr(x_i = 0) = (1-\theta)$. Obtain confidence intervals for $\theta$ by Armitage's (1958) method, and compare these with the confidence intervals obtained if the same point had been reached on a fixed sample size plan.

9. Examine how Amster's (1963) modified Bayes stopping rule could be applied to Ex. 8.4.

10. Set out a series of investigations, including empirical sampling trials, to study the effect of variations in the loss functions used for a decision theory approach to sequential estimation.

11. A random variable $X$ has a rectangular distribution of unit length but unknown mean $\theta$. The prior distribution of $\theta$ is uniform over the whole real line. The loss of estimating $\theta$ by $\theta^*$ is $k(\theta-\theta^*)^2$, where the cost of taking observations (assumed constant) is taken as the unit of constant.

Obtain the optimum sequential boundaries for estimating $\theta$. (See Wald, 1950, p. 164.)

CHAPTER 9

# Sequential Estimation of Points on Regression Functions

## 9.1. Introduction

So far in this book we have been concerned entirely with experiments in which it is merely the number of observations which is sequentially dependent. Consider the following example.

*Ex. 9.1.* Guttman and Guttman (1959) carried out a series of experiments to determine the time of onset of action of kinetin on the rate of division of *Paramecium candatum*. The ratio of the number of daily divisions in treated Paramecia to the number of daily division in untreated Paramecia, denoted $K/C$, could be assumed to increase monotonically with time of exposure of the treated group to kinetin, after a certain threshold exposure time, denoted $t'$. For exposure times up to the threshold value, the rate of division was not affected at all, and the experimenters were interested in estimating the threshold exposure, but they were not prepared to make any assumptions about the regression of $K/C$ on exposure time, other than the simple statements made above, which can be summarized

$$E(K/C) = f(t) \tag{9.1}$$

where

$$\begin{aligned} f(t) &= 1 \qquad (t \leqslant t') \\ &> 1 \qquad (t > t') \end{aligned}$$

and $f(t)$ is otherwise unknown.

In Ex. 9.1, the ratio $K/C$ for each day is considered as a random variable, and this quantity can be observed for any exposure time $t$ of the treated group. (The situation is similar to that discussed in § 1.1 where random variables could be observed at various levels.) The experimenters required only an estimate of $t'$, and they were not interested in determining $f(t)$ for $t > t'$. However, it is difficult to estimate $t'$ without making some assumptions about $f(t)$ for $t > t'$, and the experimenters therefore concentrated on an easier problem, which is to estimate the value of $t$, say $t_0$, at which the expectation of $K/C$ is $y_0$,

$$f(t_0) = y_0 \tag{9.2}$$

for $y_0 = 1{\cdot}10$, $1{\cdot}05$, etc. This latter problem is the type we discuss in this chapter. The efficiency of estimates of $t_0$ will clearly be a function of the levels used in the experiment. If we are not prepared to make assumptions about $f(t)$, observations should be placed close to $t_0$ if possible. If we can assume, for example, that $f(t)$ is linear beyond $t'$, the situation is more complicated (see Hald 1952, p. 550), but the variance of estimates of $t_0$ is reduced by putting observations close to $t_0$, provided there is a good estimate of the slope.

The type of estimation problem posed by the above example thus leads naturally to a sequential procedure. As the experiment progresses, and more observations become available, the increased information about $f(t)$ enables succeeding observations to be placed so as to yield more information on $t_0$.

In this chapter we discuss several strategies for placing levels in experiments such as Ex. 9.1. The methods suggested do not depend upon assuming a parametric form for $f(t)$, and so they do not employ standard regression techniques.

## 9.2. The Robbins–Monro Process

An experimenter observes random variables, $y(x)$, which have a distribution depending on the level $x$ at which the observation $y(x)$ was made. Let the expectation of $y(x)$ be $M(x)$,

$$E\{y(x)\} = M(x). \tag{9.3}$$

The experimenter has little or no knowledge of $M(x)$, and is not concerned with estimating it, but rather, he requires an estimate of the value of $x$, say $\theta$, at which the average response is $\alpha$,

$$M(\theta) = \alpha. \tag{9.4}$$

Robbins and Monro (1951) suggested the following sequential procedure to estimate $\theta$. Starting with an initial guess $x_1$, successive observations $y_r(x_r)$ are taken at levels $x_r$ chosen by the formula

$$x_{n+1} = x_n - a_n\{y_n(x_n) - \alpha\}, \tag{9.5}$$

where the sequence $a_r$, for $r = 1, 2, \ldots$, is a decreasing sequence of positive numbers tending to zero. After $n$ observations, $x_{n+1}$ is taken as an estimate of $\theta$.

For example, suppose $M(x)$ increases with $x$, and an estimate is required of the value of $x$ at which the average response $M(x)$ is zero. We put $\alpha = 0$, and use the formula

$$x_{n+1} = x_n - a_n y_n(x_n). \tag{9.6}$$

If $y_n$ is positive, the next observation is taken at a lower level $(x_{n+1} < x_n)$, and if $y_n$ is negative the next observation is taken at a higher level $(x_{n+1} > x_n)$. The changes in level decrease with $a_n$ (a suitable choice for $a_n$ is often $c/n$), and we expect the sequence $x_n$ to converge on $\theta$. Clearly some such assumption as

$$\left.\begin{aligned} M(x) > 0 \qquad &\text{for all } x > \theta \\ M(x) < 0 \qquad &\text{for all } x < \theta \end{aligned}\right\} \tag{9.7}$$

must be made to ensure the process will converge on the true value.

Three sets of problems immediately arise with this procedure. Firstly, it is of interest to examine the conditions necessary on $M(x)$, the distribution of $y(x)$ and the sequence $a_n$, for the process to converge on $\theta$. This set of problems has been covered extensively in great generality, see Derman and Sachs (1959), Sachs (1958) and references for some of this work. (A Bibliography of most of the work up to 1960 appears in the *Index to Annals of Mathematical Statistics* under the heading 'Stochastic approximation'.)

Secondly, it is of interest to examine the asymptotic distribution of $x_n$, and allied with this is the problem of determining the sequence $a_n$ which gives the most rapid convergence to $\theta$, and the distribution of $x_n$ having the smallest variance. This set of problems has been discussed by Chung (1954), Hodges and Lehmann (1956), Sachs (1958) and others. Thirdly, there are the problems of determining a stopping rule, and of associating confidence intervals or standard errors with the estimate of $\theta$. This third set of problems has scarcely been touched upon.

Some considerable effort has been put into obtaining multivariate generalizations of the procedure, or to making slight modifications to the process. However, in spite of all this attention, the Robbins–Monro process remains almost entirely of theoretical interest only.

One generalization which is of some practical importance is to distinguish between the Robbins–Monro process as a method of placing levels, and the estimation procedure adopted. If different estimation procedures are allowed, the up-and-down type rules discussed in § 10.3–§ 10.6 are of this type.

Even if it were desirable to survey the detailed results in this field, this would be outside the scope of the present volume. However, some simple (though restrictive) results obtained by Hodges and Lehmann (1956) do give considerable insight into the behaviour of the process. The method used by these authors is given below, with some added discussion.

Let us assume that

$$\begin{aligned} M(x) &= \beta x \\ V\{y(x)\} &= \sigma^2 \end{aligned} \tag{9.8}$$

and that the Robbins–Monro process is being run to estimate the value of $x$ for which $M(x) = 0$, which has a true value of zero. Equation (9.6) is appropriate for this situation, which we write

$$x_{n+1}(y_n, x_n) = x_n - a_n y_n(x_n) \tag{9.9}$$

to denote that $x_{n+1}$ is a function of $y_n$ and $x_n$. By taking expectations with respect to $y_n(x_n)$ conditional on $x_n$, we have

$$E(x_{n+1}|x_n) = x_n(1-a_n\beta),$$

so that

$$E(x_{n+1}) = x_1 \prod_{r=1}^{n} (1-a_r\beta), \tag{9.10}$$

which is the expected bias at $(n+1)$. We notice that this result is not always too useful, since if by chance $x_1 = 0$, the expected bias is zero for all $n$, but up to a certain sample size the distribution of $x_n$ is bimodal. However, equation (9.10) does give a measure of how far down the sequence an inappropriate choice of $x_1$ biases results.

By squaring (9.9) and taking expectations with respect to $y_n(x_n)$, conditional on $x_n$, we have

$$E\{x_{n+1}^2(y_n, x_n)|x_n\} = (1-\beta a_n)^2 x_n^2 + a_n^2\sigma^2. \tag{9.11}$$

Hence we obtain

$$E(x_{n+1}^2) = x_1^2\left\{\prod_{r=1}^{n}(1-\beta a_r)\right\}^2 + \sigma^2 \sum_{r=1}^{n} a_r^2 \prod_{s=r+1}^{n}(1-\beta a_s)^2. \tag{9.12}$$

Chung established that the constants $a_n = c/n$ give most rapid convergence of $x_n$ to $\theta$ under certain conditions, and inserting this we have

$$E(x_{n+1}^2) = x_1^2\,\phi_n^2(c\beta) + \frac{\sigma^2}{\beta^2}\psi_n(c\beta), \tag{9.13}$$

where

$$\phi_n(z) = \frac{1}{n!}\prod_1^n (r-z) = \frac{\Gamma(n+1-z)}{\Gamma(1-z)\,.\,n!} \tag{9.14}$$

and

$$\psi_n(z) = \sum_{r=1}^{n} \frac{z^2}{r^2} \prod_{s=r+1}^{n} (1-z/r)^2 = \frac{\Gamma^2(n+1-z)}{(n!)^2} \sum_{r=1}^{n} \frac{z^2(r!)^2}{r^2\,\Gamma^2(1+r-z)}. \tag{9.15}$$

The first term on the right of (9.13) arises out of bias in $x_{n+1}$, and vanishes when $c\beta$ is integral provided $n \geqslant c\beta$. It follows that asymptotically,

$$\phi_n(z) \simeq \frac{(1-z/n)^{1/2}}{n^z\,\Gamma(1-z)} \tag{9.16}$$

and the contribution of the bias term to the expected squared error is of the order $O(n^{-2c\beta})$.

The analysis of the second term, $\psi_n(z)$, is rather complicated. Consider first an approximation to

$$\left\{\frac{\Gamma(n+1-z)}{\Gamma(n+1)}\right\}^2.$$

By inserting Stirling's theorem we have

$$\left\{\frac{\Gamma(n+1-z)}{\Gamma(n+1)}\right\}^2 \simeq (1-z/n)\,(n-z)^{-2z} \simeq \{n(n-z)^{2z-1}\}^{-1}.$$

This approximation can be used in the summation part of $\psi_n(z)$. Write

$$S(z) = \sum_{r=1}^{n} \frac{1}{r^2}\left\{\frac{\Gamma(r+1)}{\Gamma(r+1-z)}\right\}^2$$

$$\simeq \sum_{r=1}^{n} \frac{1}{r^2}\,.\,r(r-z)^{2z-1} = \sum_{r=1}^{n} \frac{(r-z)^{2z-1}}{r}.$$

If $z = \frac{1}{2}$, then this last summation is asymptotically equal to $\log_e n$.

For $z > \frac{1}{2}$, the asymptotic value can be found by integrating the term under the summation with respect to $r$.

$$S(z) \simeq \int_1^n \frac{(r-z)^{2z-1}}{r} \mathrm{d}r$$

$$\simeq \int_1^n \{(r-z)^{2z-2} - z(r-z)^{2z-3} + \ldots\} \mathrm{d}r$$

$$\simeq \frac{(n-z)^{2z-1}}{2z-1}.$$

Thus we obtain

$$E(x_{n+1}^2) = \begin{cases} \sigma^2 \log_e n/(4\beta^2 n) & (c\beta = \frac{1}{2}) \quad (9.17) \\ \dfrac{\sigma^2 c^2}{n(2c\beta - 1)} & (c\beta > \frac{1}{2}). \quad (9.18) \end{cases}$$

(The above argument can be made rigorous to obtain (9.17) and (9.18) as asymptotic approximations.)

The results (9.10), (9.17) and (9.18) show that, under certain conditions on $c\beta$, the bias arising from bad initial guesses rapidly tends to zero with increasing sample size, and the expected squared error of $x_n$ is of the order $O(1/n)$.

If $M(x)$ is a more general curve than (9.8), then providing it is single valued and approximately linear near $\theta$ given by (9.4), we expect (9.18) to hold asymptotically, with $M'(\theta)$ substituted for $\beta$. Sachs (1958) proved under very general conditions that when the Robbins–Monro process is used with the constants $a_n = c/n$, the distribution of $(x_n - \theta)$ is asymptotically normal with mean zero and variance (9.18), with $M'(\theta)$ substituted for $\beta$. Other versions of the theorem have been given by Chung (1954), and Hodges and Lehmann (1956).

If the constants $a_n$ are not of the form $c/n$, the main results relate to convergence in probability and in mean square of $x_n$ to $\theta$. One of the most general forms of convergence theorem is given by Dvoretzky (1956) (see Wolfowitz (1956) and Derman and Sachs (1959) for other proofs of this theorem, and see Ventner (1966) for a generalization). The next section contains a statement and discussion of Dvoretzky's theorem, based mainly on Derman (1956). Readers not interested in the theorem should pass directly to § 9.4.

## 9.3. Dvoretzky's Theorem

The Robbins–Monro procedure is viewed as a deterministic process (which is shown to be convergent) with a superimposed random element. The Robbins–Monro process is defined by

$$x_{n+1} = x_n - a_n\{y_n(x_n) - \alpha\}.$$

Write this as

$$x_{n+1} = x_n - a_n\{M(x_n) - \alpha\} - a_n Z(x_n), \tag{9.19}$$

where

$$Z(x_n) = y_n(x_n) - M(x_n).$$

Rewrite equation (9.19) as

$$x_{n+1} = T_n(x_1, \ldots, x_n) - W_n(x_n) \tag{9.20}$$

where

$$T_n(x_1, \ldots, x_n) = x_n - a_n\{M(x_n) - \alpha\}$$

and

$$W_n(x_n) = a_n\{y_n(x_n) - M(x_n)\}.$$

If the random elements $W_n$ are suppressed, we have a deterministic process which can be shown to be convergent under certain conditions on $T_n$. Suppose we have

$$|T_n(x_1, \ldots, x_n) - \theta| < F_n|(x_n - \theta)|. \tag{9.21}$$

Then if the $W_n$ are all zero,

$$\begin{aligned} |x_{n+1} - \theta| &\leqslant |T_n(x_1, \ldots, x_n) - \theta| \\ &\leqslant F_n|(x_n - \theta)| \\ &\leqslant F_n . F_{n-1}|(x_{n-1} - \theta)| \\ &\leqslant |(x_1 - \theta)| \prod_{i=1}^{n} F_i. \end{aligned}$$

Thus if

$$\prod_{n=1}^{\infty} F_n = 0 \tag{9.22}$$

then $x_n$ converges to $\theta$.

Now suppose the $W_n$ are random variables with zero expectation and variance $\sigma_n^2$, and suppose

$$\sum_{n=1}^{\infty} \sigma_n^2 < \infty. \tag{9.23}$$

Put

$$V_n^2 = E(x_n - \theta)^2$$

then from (9.20) and (9.21)

$$V_{n+1}^2 \leqslant F_n^2 V_n^2 + \sigma_n^2$$

hence we have

$$V_{n+1}^2 \leqslant \sum_{i=1}^{n-1} \sigma_i^2 b_{n-i} + \sigma_n^2 + V_1^2 b_{n-1} \tag{9.24}$$

where

$$b_{n-i} = \prod_{r=i+1}^{n} F_r^2.$$

If we assume that $V_1^2 < \infty$, it is readily seen that the right-hand side of (9.24), and hence the left-hand side, converges to zero as $n \to \infty$. That is,

$$\lim_{n\to\infty} \{E(X_n - \theta)^2\} = 0.$$

Condition (9.21) is a form given by Derman (1956), and it is used above for ease of presentation. Dvoretzky assumed

$$|T_n(x_1, \ldots, x_n) - \theta| < \max\{\alpha_n, (1+\beta_n)|x_n - \theta| - \gamma_n\} \tag{9.25}$$

where $\alpha_n$, $\beta_n$, $\gamma_n$ are positive numbers satisfying

$$\left.\begin{aligned} \lim_{n\to\infty} \alpha_n &= 0 \\ \sum_{r=1}^{\infty} \beta_r &< \infty \\ \sum_{r=1}^{\infty} \gamma_r &= \infty \end{aligned}\right\}. \tag{9.26}$$

For a full proof of the convergence theorem the reader is referred to the references quoted above. However, the discussion given here should enable him to understand the following statement of Dvoretzky's theorem.

*Theorem 9.1* (Dvoretzky). Let $x_1$ be a random variable for which $E\{x_1^2\} < \infty$, and let $W_i$, $i = 1, 2, \ldots, n$, be random variables satisfying (9.23), and

$$E(W_n | x_1, \ldots, x_n) = 0.$$

Let $T_n(x_1, \ldots, x_n)$ be a function satisfying (9.25) with (9.26). Then the sequence of $x_r$ defined by (9.20) satisfies

$$\lim_{n\to\infty} E\{(x_n-\theta)^2\} = 0$$

and

$$\Pr\{\lim_{n\to\infty} x_n = \theta\} = 1.$$

## 9.4. Practical Use of the Robbins–Monro Process

The important feature of Sachs's and Dvoretzky's theorems is that they give a theoretical basis for employing the Robbins–Monro process under a very wide range of conditions. For most practical applications the process would be used in the form

$$x_{n+1} = x_n - \frac{c}{n}\{y_n(x_n) - \alpha\}.$$

Sachs's theorem sets out some very general conditions under which $x_n$ is approximately normal with a variance given by (9.18). A suitable choice for $c$ is to choose it to minimize (9.18). This leads to choosing $c = \beta^{-1}$, whence

$$V(x_n) = \frac{\sigma^2}{n\beta^2} \tag{9.27}$$

asymptotically. If we assume that $y(x)$ is normally distributed with mean $\beta(x-\theta)$ and variance $\sigma^2$, then we have

$$E\left\{\frac{\partial \log f(x,\theta)}{\partial\theta}\right\}^2 = \frac{\beta^2}{\sigma^2},$$

and the Cramer–Rao lower bound to the variance of estimates of $\theta$ is (9.27). A more general result holds. The Robbins–Monro process, therefore, can be fully efficient (asymptotically) if $c = \beta^{-1}$, although the practical difficulty arises that $\beta$ is unknown and must be guessed. However, the efficiency of the Robbins–Monro process is not too strongly dependent on a correct choice of $c$. The (asymptotic) efficiency is (9.27) divided by (9.18), which is $(2c\beta-1)/(c\beta)$, and some values of this ratio are given in Table 9.1. The zero asymptotic efficiency at $z = \frac{1}{2}$ is obtained because, from (9.17), the efficiency is

$$\left(\sigma^2/n\beta^2\right)\Big/\left(\frac{\sigma^2 \log_e n}{4n\beta^2}\right) = \frac{4}{\log_e n}, \tag{9.28}$$

or more precisely $4/(\log_e n+\gamma)$, where $\gamma$ is Euler's constant. The tendency to zero is therefore very gradual, and Wetherill (1963) reported some empirical sampling trials in which an efficiency of about 80% for $c\beta = \frac{1}{2}$ was observed at $n = 50$.

TABLE 9.1. Asymptotic efficiency of the Robbins–Monro process as a function of $z = c\beta$†

| $z$ | 0·50 | 0·75 | 1·00 | 1·25 | 1·50 | 2·00 | 2·50 |
|---|---|---|---|---|---|---|---|
| Efficiency | 0 | 0·88 | 1·00 | 0·96 | 0·88 | 0·75 | 0·64 |

† Extracted from Wetherill (1963).

Table 9.1 shows that there is a large range of values of $c$ for which the process is very efficient. For practical use of the strategy, some knowledge of $\beta$ will be required, in order to estimate the standard error of estimates $\hat{\theta}$, or to determine approximately how many observations are needed to achieve a specified standard error.

Assume that the variables $y_i$ are independently normally distributed with

$$E\{y(x)\} = M(x) = \beta(x-\theta)$$
$$V\{y(x)\} = \sigma^2,$$

and suppose $\theta$ and $\sigma^2$ are known, but not $\beta$. If maximum likelihood estimation is carried out for $\beta$, then by standard regression theory, the variance of estimates of $\beta$ is

$$\sigma^2/\sum_1^n (x_n-\theta)^2$$

and the contribution to the information on $\beta$ from the $n$th observation is

$$\frac{1}{\sigma^2}(x_n-\theta)^2,$$

which is of order $O(1/n)$ by Theorem 9.1. The total information on $\beta$ is approximately proportional to $\log_e n$. This contrasts markedly with the fact that each observation contributes a constant amount of information on $\theta$.

The strategy therefore provides very little information about $\beta$, except in the first few observations. The Robbins–Monro process cannot be criticized for something it was not designed to do, but an estimate effectively without a standard error is rarely of use to a statistician.

Two difficulties arise in attempting to apply the Robbins–Monro procedure to a practical problem. Firstly, observations must be taken serially and a calculation performed in between each one, which is not always convenient. Secondly, it is nearly always impracticable to stick to step sizes of $c/n$. It also appears that, the asymptotic results notwithstanding, some serious biases can arise in small samples; see § 10.2. A number of modifications to the Robbins–Monro procedure have been proposed, for which see the literature.

A great deal of work has been done recently on the application of stochastic approximation procedures such as those of Robbins–Monro and Kiefer–Wolfowitz (see § 9.5 below) to the fields of acoustics, automation and electrical engineering. For example, the techniques have been used in connection with mathematical models of learning (Chien and Fu, 1967; Fu, 1967), with automatic systems for use in sonar detection, seismic detection, and radio communications (Chang and Tuteur, 1971), and also with pattern recognition procedures. Some of this work is theoretical, rather than applied in nature, and all of it is highly technical. Access to the literature may be obtained through a standard science citation index to Robbins and Monro (1951), Dvoretzky (1956), and Kiefer and Wolfowitz (1952).

## 9.5. The Kiefer–Wolfowitz Procedure

A problem of some practical importance with a regression function $y(x)$ is to estimate the value of $x$, say $\theta$, at which the expectation of $y(x)$ is a maximum.

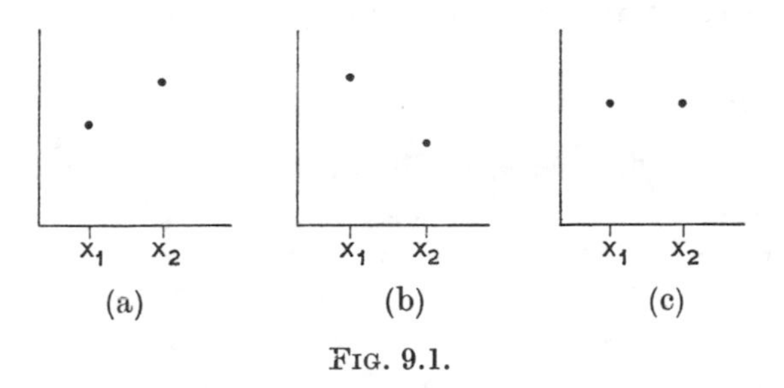


FIG. 9.1.

Suppose two observations $y_1(x_1)$ and $y_2(x_2)$ are taken at values $x_1$

and $x_2$ where $x_2 > x_1$. If $y_2 > y_1$, then we expect the maximum to be at $x > x_2$, and $x < x_1$ if $y_2 < y_1$ (see Figs. 9.1a and 9.1b). If $y_1 \simeq y_2$, there are a number of possibilities, and more observations should be taken before deciding what to do. Thus it would be reasonable to take further observations in the direction indicated by the slope of the $y$, and to take the distance of further observations from $x_1$ and $x_2$ to be proportional to the difference between $y_1$ and $y_2$. Thus, given certain conditions on $M(x) = E\{y(x)\}$, and $V\{y(x)\}$, an experimenter will have no difficulty in placing further observations $x_i$ so as to converge on $\theta$.

Kiefer and Wolfowitz (1952) defined the following procedure for stochastic estimation of the maximum of a regression function.

Let $y(x)$ be a random variable with a distribution $H(y|x)$ depending on $x$, and with an expectation

$$M(x) = \int y \, \mathrm{d}H(y|x). \tag{9.29}$$

Suppose that the variance of $y$ is finite for all $x$,

$$\int \{y - M(x)\}^2 \mathrm{d}H(y|x) = V\{y(x)\} \leqslant S < \infty.$$

Suppose further that $M(x)$ is strictly increasing for $x < \theta$, and strictly decreasing for $x > \theta$. Define infinite sequences of positive numbers, $a_n$, $c_n$, such that

$$\begin{aligned} c_n &\to 0 \\ \sum a_n &= \infty \\ \sum a_n c_n &< \infty \\ \sum a_n^2 c_n^{-2} &< \infty. \end{aligned}$$

For example, $a_n = 1/n$, $c_n = 1/n^{1/3}$.

Define a recursive scheme of estimators $z_i$ of the maximum,

$$z_{n+1} = z_n + \frac{a_n(y_{2n} - y_{2n-1})}{c_n} \tag{9.30}$$

where $y_{2n}$ has a distribution $H(y|z_n + c_n)$ and $y_{2n-1}$ has a distribution $H(y|z_n - c_n)$.

To operate this recursive scheme, two observations are necessary at each stage, and the distance and direction of movement from any point $z_r$ depend only on the two observations in the $r$th stage. The scheme follows the reasoning laid down above, given with reference to Fig. 9.1.

This scheme is strictly defined, and various situations might cause trouble which would give no difficulty to an experimenter who proceeds in an *ad hoc* way from the results before him. For example, if initial observations are made near the tails of a bell-shaped curve, convergence to the maximum point by (9.30) would be incredibly slow. Similarly, if two observations were by chance very different, formula (9.30) would make a large adjustment which could put the next pair of observations well away from the maximum. A human experimenter could probably deal successfully with a far greater range of conditions than those under which the Kiefer–Wolfowitz process can be expected to converge. Further, a human experimenter can use all the observations, and not just the last two, in deciding movement from a given point. Even if a parametric formulation is assumed, the Robbins–Monro process is very highly efficient, but the Kiefer–Wolfowitz process is not efficient in the parametric case, see the discussion below and P9.4. However, the problem of determining the maximum of a regression function is of considerable practical importance, and any theoretical investigations which may throw light on the problem are of some value.

It is very difficult to obtain any measure with which to compare the Kiefer–Wolfowitz procedure. In practice many statisticians would probably proceed as follows. Take a number of observations at three points $x_1$, $x_2$ and $x_3$ and fit a quadratic curve to the results. A second, and possibly a third, stage could now be used, with observations placed with the help of the results from the first stage.

Suppose we can assume that $E\{y(x)\}$ is a quadratic curve with a maximum at $x = \alpha$, and

$$\begin{aligned} E\{y(x)\} &= \gamma - \beta(x-\alpha)^2 \\ V\{y(x)\} &= \sigma^2, \end{aligned} \tag{9.31}$$

where $\gamma$, $\beta$, $\alpha$ are unknown, but $\sigma^2$ is known. If the $y$ are normally distributed, the log-likelihood is proportional to

$$S = -\sum \{y_i(x_i) - \gamma + \beta(x_i - \alpha)^2\}^2. \tag{9.32}$$

Suppose three groups of observations are taken, $n$ at $x_2$, and $\frac{1}{2}(N-n)$ at $x_1 = x_2 - d$, and at $x_3 = x_2 + d$. If the centre group is placed exactly at $\alpha$, then by differentiation of (9.32) we find that the asymptotic variance of $\hat{a}$ is

$$V(\hat{a}) = 1/4\beta^2 d^2(N-n). \tag{9.33}$$

This result shows that for optimum estimation of $\alpha$, the observations must be put as far away from $\alpha$ as possible.

If the maximum $\alpha$ is outside the range of the $x$ used, it can be shown that a similar result holds, but with a distribution of observations having $n = \frac{1}{2}N$ at the centre group. The final variance of $\hat{a}$ is proportional to $1/N$, but can be made indefinitely small by placing the groups far enough apart. This optimum placing of observations depends critically on the response curve being quadratic (9.31), and so is not of much use in practical situations, where (9.31) would usually be an approximation valid only in the neighbourhood of $\alpha$. However, the contrast between these results and the Kiefer–Wolfowitz procedure is very marked; for the Kiefer–Wolfowitz procedure the observations themselves converge on the maximum, since the constants $c_n$ are chosen to tend to zero.

Burkholder (1956) proved that under certain conditions, the Kiefer–Wolfowitz procedure can still be used if $c_n$ is held constant for all $n$ at a particular value $c_0$, and $z_n$ is then asymptotically normally distributed with a variance proportional to $1/n$. This result is difficult to use since in practice there will rarely be enough information about the response curve to choose $c_0$.

Sachs (1958) has discussed conditions for asymptotic normality of $z_n$. If $c_n$ is chosen to tend to zero, the asymptotic variance of $z_n$ can never be made as small in order of magnitude as Burkholder's result of $o(1/n)$. The discussion and conditions given in Sachs's paper are involved, and will not be reproduced here.

The convergence in probability and in mean square of the Kiefer–Wolfowitz procedure follow from Dvoretzky's theorem.

For work on improvements to the Kiefer–Wolfowitz procedure see Fabian (1967, 1969).

## 9.6. Methods for Seeking the Maximum of a Response Surface

In Ex. 1.1, the yield $y$ of a chemical process depended on a series of variables $x_1, x_2, \ldots, x_k$, defining the pressure, temperature and other conditions of the process. It is assumed that little is known about the response surface $E\{y(x_1,\ldots,x_k)\}$, and that a series of experiments is to be performed to estimate the value or values of $(x_1,\ldots,x_k)$ for which $E\{y(x_1,\ldots,x_k)\}$ is a maximum. A multivariate form of the Kiefer–Wolfowitz process would be applicable (see Blum, 1954), but other methods are preferred for both practical and theoretical reasons. The difficulty in analysing and comparing these alternative methods lies in that they are not precisely formulated; we shall not study them in detail in this volume, but instead we shall briefly describe the main methods and list references for further reading.

*(a) Non-sequential method*

One possible approach would be to use one big factorial experiment, using several levels of each variable $x_i$, and explore the whole region in which the maximum is thought to lie. A quadratic surface could be fitted in the neighbourhood of the maximum. The disadvantages of this method are that a large number of observations will usually be involved, many of which will provide a negligible amount of information on the position of the maximum. There is also a danger that the entire experiment is wrongly placed and does not include the maximum, or that observations are too far apart and give too flat an impression of the surface. In experiments such as agricultural trials, where each observation takes a long time to obtain, a plan basically similar to this one must be used, but otherwise methods (*c*) or (*d*) below would normally be preferable.

*(b) Random strategy*

Another approach is to define the area of the space of the $x$ to be explored, and then take observations at randomly selected points, perhaps using stratified random sampling. The possibilities of random strategies do not seem to have been fully explored; they were suggested by Anderson (1953) and briefly studied by Brooks (1958, 1959). One suggestion is merely to take the winner of the randomly selected points as an estimate of the maximum. However, it would probably be better to use two or more stages. At the first stage, the space to be explored is stratified and a randomly selected point in each observed. At the second stage, further randomly selected observations would be taken in the strata giving highest results.

*(c) One-at-a-time*

This method is to explore only in one $x_i$ at a time, holding all the other variables fixed, and find the maximum for these conditions. Then another variable is varied, and so on, until the maximum appears to have been reached. Often two or three complete rounds of all $x_i$ will be necessary to converge sufficiently on the maximum. The first exposition of the one-at-a-time method is given by Friedman and

Savage (1947). The univariate method can be very wasteful in observations with certain types of response surface, as with a ridge, for example. Further, the number of observations required tends to rise fairly rapidly with the number of factors. However, in a small number of dimensions the method often works well.

(*d*) *Steepest ascent*

Box and Wilson (1951) developed a strategy which has received much attention in the literature (see references). A small set of observations are taken centred on a guess of the maximum, and a plane surface is fitted to the results. The direction in which the yield is increasing at maximum rate is found, and a further small series of observations is taken some distance from the first set in this direction (the direction of steepest ascent). Thus if the first observations are centred on $(x_1, x_2, \ldots, x_k)$, the vector of changes to the $x$ is found for which $E\{y(x_1, \ldots, x_k)\}$ is increasing most rapidly. New observations are taken in this direction and the process repeated.

When there is evidence that the region of the maximum has been reached, a new set of observations is taken, a quadratic surface fitted, and the maximum estimated.

The two stages of the steepest ascent method require different experimental designs. The first stage involves designs suitable for estimating the direction of steepest ascent. The second stage involves a design suitable for estimating a quadratic surface. The design aspect of these problems has received considerable attention, but there are other problems more difficult to study, such as how much effort to put into experiments at the first stage, how far to move the centre of first stage experiments, when to change to designs suitable for fitting a quadratic surface, etc.

A good description of the steepest ascent method, with numerical illustration and references, is given in Duncan (1959), and Davies (1956). A bibliography of work on response surface designs appears under that heading in Herzberg and Cox (1969).

*Comparison of methods*

Brooks (1959) gives a comparative study of the maximum seeking methods listed here, using two dimensions and four different surfaces. His measure of effectiveness for any method was the average response at the estimate of the maximum. Only a few trials were run for each strategy, but the results show the steepest ascent method to be best, and the one-at-a-time method next best. More extensive trials with greater numbers of factors would be of interest.

It would appear possible to adopt a Bayesian approach to maximum seeking methods. A prior distribution for the position of the maximum point could be assumed, and it would be necessary to assume a form for the response surface. The basic difficulty in a Bayesian approach to problems of this kind is to find a suitable formulation. Even if the approach is rather academic, it may suggest new strategies, or serve as a model for evaluating existing ones.

Anderson (1953), gives a survey of maximum seeking methods up to 1953, and Duncan (1959) has some general discussion, as well as the description of the steepest ascent method already referred to. The subject is of great practical interest, and for discussions relating to applications see Box (1954).

A related but slightly different problem is that of maximizing the yield, or perhaps some other quantity, of a manufacturing process using the full scale process itself, and not a pilot scale process for experimentation. The method used in this case is called EVOP, and is carefully described in Box and Draper (1969).

## Problems 9

1. Assume the simple model of (9.8) for the response $y(x)$ at a level $x$, and assume further that the responses are independently and normally distributed. Examine the amount of information given by the Robbins–Monro process on the slope $\beta$ of the response curve.

2. Suppose the response $y(x)$ has the form

$$M\{y(x)\} = M(x) = \alpha\{1+e^{-\beta(x-\mu)}\}^{-1}$$
$$V\{y(x)\} = \sigma^2,$$

where $\alpha$, $\beta$,$\mu$ and $\sigma^2$ are unknown, and the Robbins–Monro process is being used to estimate points $M(\mu) = \alpha/2$. Suppose that the steps in $x$ used are small in comparison with $1/\beta$. What biases arise purely out of the curvature of the response curve?

3. Verify the application of Dvoretzky's theorem to the Robbins–Monro and Kiefer–Wolfowitz processes.

4. In the Kiefer–Wolfowitz process, we can distinguish between the

strategy as a method of concentrating observations in the desired region, and the estimation rule adopted. Critically discuss the estimation rule, and obtain the asymptotic efficiency in comparison with maximum likelihood estimates of $\alpha$, if the model (9.31) was true and the random variables were normally distributed. What objections are there to a parametric estimation procedure?

5. Consider the experiment discussed by Guttman and Guttman (1959) described in Ex. 9.1, and examine estimation and design problems if it could be assumed either that

$$M(x) = \begin{cases} \alpha, & (x \leqslant \theta) \\ \alpha + \beta(x-\theta) & (x \geqslant \theta) \end{cases}$$

or that

$$M(x) = \begin{cases} \alpha & (x \leqslant \theta) \\ \alpha + \beta(x-\theta) + \beta_2(x-\theta)^2 + \ldots & (x \geqslant \theta) \end{cases}$$

where the power series defines a single-valued monotonic increasing function of $x$. (Estimates of $\theta$ are required, but none of the parameters is known. Assume the variables to be normally distributed with unknown but constant variance.)

Set up a simple multi-stage procedure to estimate $\theta$. Design a set of sampling trials to compare the following three strategies for estimation of $\theta$:

(a) The Robbins–Monro process.

(b) The multi-stage procedure you have set up.

(c) Subjective judgment. Explain the experiment to a subject, and allow him to judge where to put the next observation. (You will have to set a starting value, and give some rough information on the ranges of $x$ the subject has to explore.)

6. When the Robbins–Monro process is being used to estimate a point on a regression curve such as in the simple model discussed in § 9.2, equation (9.8), any confidence interval statement will depend on the unknown slope $\beta$, for which the process provides a very poor estimate. Consider the problem of setting up a strategy for obtaining interval estimates of $\theta$ for which $M(\theta) = \alpha$. For normally distributed random variables, discuss the asymptotic efficiency you would aim to achieve.

CHAPTER 10

# Sequential Estimation of Points on Quantal Response Curves

## 10.1. Introduction

Consider the following examples.

*Ex. 10.1.* Some 1-ft lengths of plastic conduit pipe are subjected to impacts of various energies. Several features of the response could be noted, but one of particular importance is to record whether or not a brittle failure has occurred.

*Ex. 10.2.* A subject is supplied with a pair of headphones, and a short note is sounded through them, with a time interval $t$ between the note sounding at the left and right earphones. From the subject's point of view, the note appears to originate from within his head, somewhere on a line joining the two ears. The effect of varying $t$, to the subject, is to make the note seem to originate from a different part of this line. The subject is made to guess, after each stimulus, whether the note seemed to come from the right or left side of his head.

For both of these examples, a statistical model of the following type is appropriate. A stimulus can be applied at various levels $x$, and the response $y(x)$ is a random variable taking on two values, one and zero, the probability of a positive response being $P(x)$. For increasing $x$, $P(x)$ increases gradually from zero to unity, following some S-shaped curves such as the logistic curve,

$$P(x) = \{1+e^{-\gamma(x-\alpha)}\}^{-1}. \tag{10.1}$$

Thus in Ex. 10.1, $P(x)$ is the probability of a brittle failure, which increases from zero to one as the energy of impact increases. For Ex. 10.2 $P(x)$ is the probability of the subject's judging left (or alternatively, right), which increases from zero to unity as the time interval is varied.

For response curves of this type, an experimenter must first decide the region or characteristic of interest. For Ex. 10.1, one of the most important features is the location of the lower end of the response curve. This can be defined in terms of the energy level, $L_p$, for which the probability of a positive response is $p$,

$$P(L_p) = p$$

where $p$ is perhaps 10% or 15%.

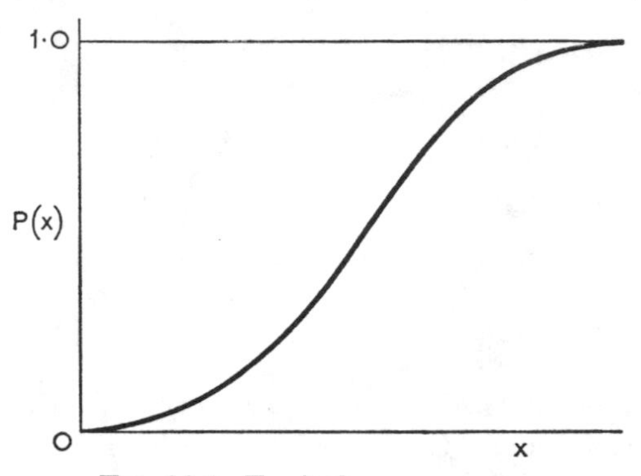


FIG. 10.1. Typical response curve.

For Ex. 10.2 it is not the location, but the slope of the response curve which is of principal interest. For the logit model (10.1), it can be shown that estimation of $\gamma$ is most efficient if the observations are placed in two groups, near the $L_{0\cdot 09}$ and $L_{0\cdot 91}$ levels. (For various reasons, less extreme values were used in practice.)

Questions concerning response curves can often be put in terms of the problem of estimating levels $L_p$, and in this chapter some strategies are discussed which were aimed at providing simple and efficient point estimates of levels $L_p$. (There is an associated problem of providing interval estimates of levels $L_p$, and there has so far been no progress on this.)

### 10.2. The Robbins–Monro Process

If the conditions of Theorem 9.1 are examined, it will be found that they include the quantal response set-up, such as the logit quantal response model (10.1). Equation (9.5) becomes

$$x_{n+1} = x_n - \frac{c}{n}\{y_n(x_n) - p\}, \tag{10.2}$$

and after $n$ observations, $x_{n+1}$ is used as an estimate of $L_p$. The asymptotic variance depends on the derivative of the response curve $P(x)$ at $L_p$, which is denoted

$$\beta = \left\{\frac{\mathrm{d}P(x)}{\mathrm{d}x}\right\}_{x=L_p}.$$

The variance of $x_n$ is asymptotically

$$\frac{pqc^2}{n(2c\beta-1)}$$

from (9.18), where $q = 1-p$ provided $c\beta > \frac{1}{2}$. This variance takes a minimum for choice of $c$ such that $c\beta = 1$, when the variance becomes

$$V(x_n) = (n\gamma^2 pq)^{-1}. \tag{10.3}$$

This is exactly the lower bound to the asymptotic variance of estimators of $L_p$, for all $p$. In fact, much of the previous discussion on the Robbins–Monro process holds here for the quantal response set-up.

Small sample empirical sampling trials were run by Wetherill (1963) – see also Davis (1963). Some of Wetherill's results are reproduced in Table 10.1. For estimation of $L_{0\cdot50}$, the behaviour of the Robbins–Monro process is in very good agreement with asymptotic theory, even for samples of size 35. However, away from the immediate neighbourhood of $L_{0\cdot50}$, the process leads to small sample estimates which frequently have large biases, and in addition, the sample variances are greatly in excess of those predicted from asymptotic theory.

The main reason for this behaviour is the following. Suppose an experimenter wishes to estimate $L_{0\cdot90}$, and that his initial level $x_1$ is very close to the true value. Suppose further that the first observation is zero (as it will be once in every ten trials), then the second observation will be taken at the level

$$x_2 = x_1 - c(0-0\cdot90) = x_1 + 0\cdot90c.$$

This value, $x_2$, may well be far above $L_{0\cdot90}$, so that the second observation will nearly always be positive, leading to $x_3$ at

$$x_3 = x_2 - \frac{c}{2}(1-0\cdot90) = x_1 + 0\cdot85c.$$

TABLE 10.1. Empirical sampling trials of the Robbins–Monro process. Logit quantal model (10.1), $\alpha = 0, \gamma = 1$†

(a) Estimation of $L_{0\cdot50}$ (500 runs in each set)

| | Average squared error | | |
|---|---|---|---|
| *Sample size 25* | | $c$ | |
| Starting value | 2·5 | 4·0 | 5·5 |
| 0 | 0·189 | 0·172 | 0·178 |
| 1·5 | 0·227 | 0·171 | 0·189 |
| Asymptotic variance | 0·250 | 0·160 | 0·172 |
| *Sample size 35* | | | |
| Starting value | | | |
| 0 | 0·146 | 0·132 | 0·126 |
| 1·5 | 0·169 | 0·143 | 0·139 |
| Asymptotic variance | 0·178 | 0·114 | 0·124 |

(b) Estimation of $L_{0\cdot75}$ and $L_{0\cdot95}$ (true values 1·10 and 2·94 respectively) 1,000 runs in each set

| Per cent. | Sample value | Starting value | $c$ | Sample mean | Sample variance | Asymptotic variance |
|---|---|---|---|---|---|---|
| 75 | 25 | 1·10 | 5·0 | 1·36 | 0·36 | 0·21 |
| 75 | 25 | 1·10 | 5·3 | 1·40 | 0·35 | 0·21 |
| 75 | 35 | 1·10 | 5·0 | 1·31 | 0·28 | 0·15 |
| 75 | 35 | 1·10 | 5·3 | 1·30 | 0·23 | 0·15 |
| 95 | 35 | 2·94 | 16·05 | 4·50 | 7·95 | 0·67 |
| 95 | 35 | 2·94 | 21·05 | 5·57 | 16·14 | 0·60 |

† Extracted from Wetherill (1963).

If $y_3$ is also positive, then $x_4$ is

$$x_4 = x_3 - \frac{c}{3}(1 - 0\cdot90) = x_1 + 0\cdot816c.$$

In fact, a minimum of about $e^{10}$ observations is necessary to pass below $x_1$.

There have been attempts to make variations on the Robbins–Monro process to deal with difficulties such as the one mentioned above, see particularly Kesten (1958), and also Davis (1963), and references, for some of this work.

## 10.3. The Up-and-Down Rule

Dixon and Mood (1948) suggested a strategy for estimating $L_{0\cdot50}$, which is much easier to operate than the Robbins–Monro process.

Choose a series of equally spaced levels of $x$; observations $y(x)$ are taken only at these values. The first observation is taken at the best guess of $L_{0\cdot50}$ available, and trials are made sequentially. For each positive response obtained, the following trial is made at the next lower level. After each zero response, the following trial is made at the next higher level. If we denote positive results by a 1, and zero results by 0, the Up-and-Down method traces out a sequence of trials on the fixed levels chosen, such as that illustrated in Fig. 10.2.

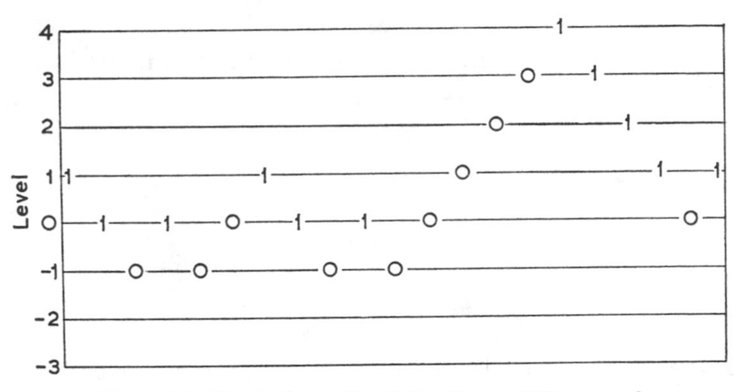


FIG. 10.2. Typical result of the Up-and-Down rule

Dixon and Mood assumed the probit model rather than the logit model (10.1), and they obtained simple approximations to the maximum likelihood estimators of $L_{0\cdot50}$ and the slope parameter $\gamma$.

A study of the asymptotic properties of the Up-and-Down rule reveals several important properties. The first step necessary is to derive the asymptotic distribution of the observations at the fixed series of levels.

Without loss of generality, the scale of measurement of the levels can be defined in terms of the distance between levels as a unit, and the starting level of the sequence, $x_1$, can be taken as the origin of the scale. Thus the $i$th level, denoted $l_i$, is written $i$ in this scale, for all $i$

integral and zero. If the true response curve is (10.1), then the probability of a positive response at level $l_i$ is

$$p_i = \{1+e^{-\gamma(i-\alpha)}\}^{-1}. \tag{10.4}$$

Let $P_i(m)$ denote the probability that the $m$th observation is taken at $l_i$. If the $m$th observation is at $l_i$, then the $(m-1)$th observation must have been at $l_{i+1}$ or $l_{i-1}$, thus $P_i(m)$ must satisfy

$$P_i(m) = P_{i+1}(m-1)\,p_{i+1}+P_{i-1}(m-1)\,(1-p_{i-1}), \tag{10.5}$$

with boundary conditions $P_0(1) = 1$, $P_i(1) = 0$, for $i \neq 0$. The asymptotic distribution of levels is given by a set of $P_i$ satisfying (10.5), which are independent of $m$.

A simplification of (10.5) for the asymptotic case arises as follows. The Up-and-Down rule is such that the number of positive results at $l_i$ never differs by more than one from the number of zero results at $l_{i-1}$. Hence asymptotically these must be equal, and the asymptotic distribution of levels is the set of $P_i$ satisfying

$$P_i p_i = P_{i-1}(1-p_{i-1})$$

or

$$P_i = P_0 \prod_{k=0}^{i-1} (1-p_k) \Big/ \prod_{j=1}^{i} p_j \qquad \text{for } i \geqslant 1 \tag{10.6}$$

and

$$P_i = P_0 \prod_{j=i+1}^{0} p_j \Big/ \prod_{k=i}^{-1} (1-p_k) \qquad \text{for } i \leqslant -1. \tag{10.7}$$

If there are $n_i$ observations at $l_i$, of which $r_i$ are positive, then the second derivatives of the log-likelihood for (10.1) are

$$\left.\begin{aligned} \frac{\partial^2 L}{\partial \alpha^2} &= -\gamma^2 \sum n_i p_i(1-p_i) \\ \frac{\partial^2 L}{\partial \alpha\, \partial \gamma} &= \gamma \sum (i-\alpha) n_i p_i (1-p_i), \\ \text{and} \quad \frac{\partial^2 L}{\partial \gamma^2} &= -\sum (i-\alpha)^2 n_i p_i(1-p_i). \end{aligned}\right\} \tag{10.8}$$

By inserting the asymptotic values $nP_i$ for $n_i$ in these equations we obtain the terms of the information matrix of $\alpha$ and $\gamma$,

$$M = -\begin{pmatrix} E\left(\dfrac{\partial^2 L}{\partial \alpha^2}\right) & E\left(\dfrac{\partial^2 L}{\partial \alpha\, \partial \gamma}\right) \\ E\left(\dfrac{\partial^2 L}{\partial \alpha\, \partial \gamma}\right) & E\left(\dfrac{\partial^2 L}{\partial \gamma^2}\right) \end{pmatrix}.$$

The inverse of this matrix gives the asymptotic variances and the covariance of maximum likelihood estimates of $\alpha$ and $\gamma$. Table 10.2 below gives some values of $nV(\hat{a})$ and $nV(\hat{\gamma})$ for the logistic model.

TABLE 10.2 Asymptotic variances of the Up-and-Down rule for the logit model, $\alpha = 0, \gamma = 1$†

| Spacing between levels | 0·25 | 0·5 | 1·0 | 1·5 | 2·0 |
|---|---|---|---|---|---|
| $nV(\hat{\alpha})$ | 4·25 | 4·50 | 5·02 | 5·57 | 6·12 |
| $nV(\hat{\gamma})$ | | 17·71 | 6·10 | 4·86 | 4·41 |

† Extracted from Wetherill (1963).

Now $\alpha$ is $L_{0\cdot50}$, and the minimum asymptotic variance for estimates of $\alpha$ is readily seen to be $4/n$. Table 10.2 shows, therefore, that if the spacing by slope product is small enough, say 0·25, then the resulting estimates of $L_{0\cdot50}$ have a very high asymptotic efficiency. Some empirical sampling trials by Wetherill (1963) demonstrate that even at $n = 35$, the asymptotic formula is a good approximation to $V(\hat{a})$ provided the sequences are started at (or near) the true $L_{0\cdot50}$, see Table 10.3.

TABLE 10.3. Empirical sampling trials on the Up-and-Down rule. Logit model, $\alpha = 0, \gamma = 1$. Maximum likelihood estimation (700 runs in each set)†

| | Average | | Sample variance | | Asymptotic variance | |
|---|---|---|---|---|---|---|
| Spacing | $L_{0\cdot50}$ | Slope | $L_{0\cdot50}$ | Slope | $L_{0\cdot50}$ | Slope |
| 0·5 | 0·038 | 1·673 | 0·119 | 0·810 | 0·129 | – |
| 1·0 | 0·014 | 1·302 | 0·128 | 0·334 | 0·143 | 0·174 |
| 2·0 | −0·012 | 1·190 | 0·171 | 0·203 | 0·175 | 0·126 |

† Extracted from Wetherill (1963).

Regarding estimation of $\gamma$ from the Up-and-Down rule, Table 10.2 shows that small spacings give very little (asymptotic) information on this. Further, the empirical sampling trials of Table 10.3 show large and variable biases in the small sample maximum likelihood estimates, and sample variances greatly in excess of the asymptotic value. See Wetherill (1963) for a fuller discussion of these points, where I conjectured that useful estimates of $\gamma$ are unlikely to be obtained from small samples with strategies of the Up-and-Down type.

Brownlee, Hodges and Rosenblatt (1953), showed that an estimator, which is asymptotically equivalent to Dixon and Mood's approximation and to the maximum likelihood estimator, is

$$\hat{L}_{0\cdot 50} = \frac{1}{n} \sum_{r=2}^{n+1} x_r \tag{10.9}$$

for a sequence of $n$ trials, where $x_r$ is the level used at the $r$th trial. (The level of the first trial gives no information, but the level which would have been used at the $(n+1)$th trial does.) Small sample properties of the average level estimator (10.9) are readily obtained.

It is convenient here to drop the convention of always starting a sequence at the origin of the scale of measurement. Write

$$S_n(l_i) = \sum_{r=2}^{n+1} x_r \tag{10.10}$$

where the first level used $(x_1)$ is at level $l_i$. Then if the first response $y_1 = 0$,

$$S_{n+1}(l_i) = l_i + 1 + S_n(l_{i+1}),$$

and if $y_1 = 1$,

$$S_{n+1}(l_i) = l_i - 1 + S_n(l_{i-1}).$$

Thus

$$E\{S_{n+1}(l_i) - \alpha\} = l_i + 1 - 2p_i + p_i E\{S_n(l_{i-1}) - \alpha\} + (1 - p_i) E\{S_n(l_{i+1}) - \alpha\}. \tag{10.11}$$

Similarly, a recurrence relation for the small sample average squared error of $S_n(l_i)$ can be obtained. Brownlee *et al.* used the probit model, rather than the logistic, but the qualitative conclusions are the same for both, and can be summarized approximately in Fig. 10.3.

Although small spacings are best asymptotically, in small samples very considerable bias can arise out of the error in the starting value. For a given starting error, the expected squared error rapidly increases as the spacing by slope product decreases, and in order to avoid paying a high price for bad starting errors, small spacings must be avoided. Usually, $\gamma$ is unknown, and only very imprecise estimates of it are available from the strategy. The spacing must be chosen on the high side to be safe, so that the strategy will generally be used for a value of spacing by slope product for which the strategy is rather inefficient.

Apart from the ease of operation and estimation, the Up-and-Down rule has little to recommend it. For a comparison of the Robbins–Monro, Up-and-Down, and other designs in the context of Bio-assays see Davis (1971). A comparison of the Up-and-Down rule, and the optimum Bayesian sequential plan for estimating $L_{0\cdot 50}$ is attempted by Freeman (1970), but he is unable to get results for designs using more than three dose levels.

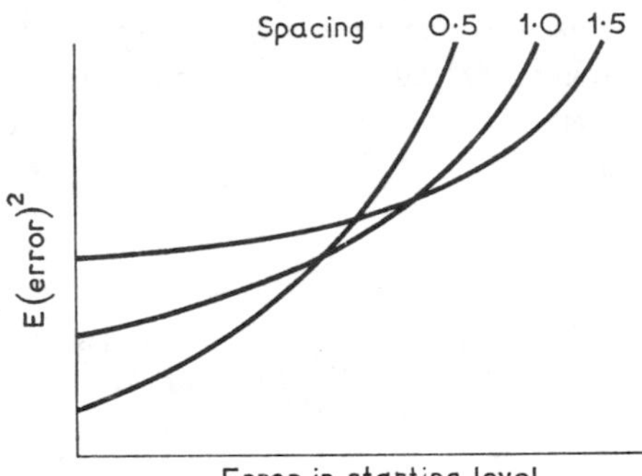


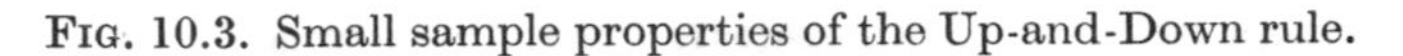
FIG. 10.3. Small sample properties of the Up-and-Down rule.

## 10.4. Modifications to the Up-and-Down Rule

The asymptotic variances given in Table 10.2 indicate that very little efficiency is gained by reducing the spacing by slope product below about 0·50. If a preliminary run with a large spacing could be used, then the sequence could be restarted with a smaller spacing, at a level close to the $L_{0\cdot 50}$ estimate obtained from the preliminary run. If the smaller spacing corresponds to a spacing by slope product of about 0·50 or less, very efficient estimates will result.

One difficulty with this suggestion is that estimates from runs of fixed length from the Up-and-Down rule vary greatly in precision from sample to sample. (For example, the run which is a sequence of positive results only gives no estimate at all.) A stopping rule is needed for the preliminary run suggested above, which will yield estimates of more uniform precision. A suitable stopping rule can be developed as follows.

The sequence of results shown in Fig. 10.2 can be considered as a sequence of ten changes of response type, the changes being at observations 3, 4, 5, 7, 9, 10, 11, 16, 20 and 21. If a sequence of observations is terminated after a fixed number of changes of response type, rather than after a fixed number of observations, there will be several advantages. Firstly, if a bad initial guess is made, and several observations are required before the first change of response type, then more observations will be taken. Secondly, in small samples, sequences giving very imprecise estimates arise mainly from those involving only one or perhaps two changes of response type, and the termination rule is easily chosen to eliminate these. Thirdly, the average number of observations per change of response type increases inversely as the spacing by slope product. Thus, if the spacing chosen is large, a relatively small number of observations is used on the (less efficient) preliminary run.

Wetherill (1963) therefore proposed the following strategy.

*Routine 11.* Proceed by the Up-and-Down rule until the $k$th change of response type. Then form an estimate of $L_{0\cdot50}$, and restart the sequence near this estimated value, using half the spacing interval used up to the $k$th change.

Some empirical sampling trials of this strategy are given in the source paper, and it is demonstrated that the strategy can be very efficient, and is less sensitive to the initial choice of spacing than the Up-and-Down rule. A suitable value for $k$ is about 6. A further division of the spacing interval could be employed, but it is doubtful whether this would be worth while in many cases. The modified strategy combines much of the efficiency of the Robbins–Monro process with the ease of operation of the Up-and-Down rule.

Lastly, I propose a modified estimation procedure. Consider each run of Fig. 10.2 separately. The first three observations form the first run, and a suitable point estimate of $L_{0\cdot50}$ from these alone would be half-way between the interval at the end of the run, at $x = -\frac{1}{2}$. Proceeding in this way, estimates of $L_{0\cdot50}$ from all ten runs are, for the five valleys, $\{-\frac{1}{2}, -\frac{1}{2}, -\frac{1}{2}, -\frac{1}{2}, +\frac{1}{2}\}$, and for the five peaks $\{-\frac{1}{2}, +\frac{1}{2}, -\frac{1}{2}, +3\frac{1}{2}, +\frac{1}{2}\}$. Denote these separate estimates of $L_{0\cdot50}$ by $w_i$, and consider the average

$$\bar{w} = \frac{1}{k} \sum w_i$$

(for $k$ changes), as a combined estimate of $L_{0\cdot50}$. Some empirical sampling trials comparing the $\bar{w}$ and average level estimates are given in Table 10.4. Further empirical sampling trials are given in Wetherill and Chen (1965). There is very good basis for using the $\bar{w}$ estimate with Routine 11, or with the Up-and-Down rule, in place of earlier suggestions for a simple estimator.

TABLE 10.4. Empirical sampling trials of the Up-and-Down rule to compare methods of estimation. (1,000 runs in each set. Each run terminates at the $k$ change of response type)†

| | | | | Estimator | |
|---|---|---|---|---|---|
| $k$ | Starting value | Spacing | Av. no. obs. | $\bar{w}$ Av. (error)$^2$ | Brownlee *et al.* Av. (error)$^2$ |
| 8 | 0 | 0·5 | 15·11 | 0·1858 | 0·2133 |
| 8 | 3 | 0·5 | 18·37 | 0·5535 | 0·5865 |
| 15 | 0 | 0·5 | 27·51 | 0·1377 | 0·1472 |
| 15 | 3 | 0·5 | 30·97 | 0·2581 | 0·2585 |
| 15 | 0 | 1·0 | 25·50 | 0·1782 | 0·1870 |
| 15 | 3 | 1·0 | 27·03 | 0·2305 | 0·2326 |

† Extracted from Wetherill and Chen (1965).

The $\bar{w}$ estimator is important in connection with a generalization of the Up-and-Down rule given in § 10.6. Some properties of the estimator can be studied by using a finite Markov chain approximation, which is discussed in the next section.

## 10.5. Finite Markov Chain Approximation

The new method of estimation suggested in § 10.4 for the Up-and-Down rule can be viewed as a Markov chain, if the state of a sequence of observations is defined only when there has been a change of response type (that is, for observations 3, 4, 5, 7, 9, etc., of Fig. 10.2). The state of the sequence is then defined by the estimator $w_i$ made at the change, coupled with the specification of peak or valley for the change.

It will be satisfactory here to truncate the levels possible, by choosing some large $k$ and putting $p_k = 1{\cdot}0$, $p_{-k} = 0$, so that there are only $(2k+1)$ levels. The set of possible states can be written

$$S' = \{S^p_k, \ldots, S^p_{-k+1}, S^v_{k-1}, \ldots, S^v_{-k}\} = \{S^p, S^v\}$$

where the suffices refer to the levels $l_i$ at which the states are defined. (Since we have put $p_k = 1{\cdot}0$, $l_k$ cannot be a valley, and similarly, $l_{-k}$ cannot be a peak.)

Given that the sequence is in a state $S^p_i$, then only valleys $\{S^v_{i-1}, \ldots, S^v_{-k}\}$ are possible as the next state. The probability of arriving at $S^v_j$ from $S^p_i$ is

$$\begin{aligned} t_{ij} &= q_j \prod_{r=j+1}^{i-1} p_r \qquad (j < i) \\ &= 0 \qquad\qquad\qquad (i \leqslant j) \end{aligned} \tag{10.12}$$

where $q_j = 1 - p_j$.

Similarly, the probability of arriving at $S^p_j$ as the next state from $S^v_i$ is

$$\begin{aligned} t_{ij} &= p_j \prod_{r=i+1}^{j-1} q_r \qquad (j > i) \\ &= 0 \qquad\qquad\qquad (i \geqslant j). \end{aligned} \tag{10.13}$$

The matrix of transition probabilities is

$$T = \left(\begin{array}{c|c} 0 & B \\ \hline A & 0 \end{array}\right) \tag{10.14}$$

where

$$A = \begin{pmatrix} t_{k,k-1} & 0 & \cdots & 0 \\ \vdots & & \ddots & \vdots \\ \vdots & & & 0 \\ t_{k,-k} & \cdots & & t_{-k+1,-k} \end{pmatrix}$$

and

$$B = \begin{pmatrix} t_{k-1,k} & \cdots & & t_{-k,k} \\ 0 & \ddots & & \vdots \\ \vdots & & \ddots & \vdots \\ 0 & \cdots & 0 & t_{-k+1,k} \end{pmatrix}$$

This Markov chain is periodic, with period 2, and there is a very strong dependence between successive $w_i$.

For a sequence started at any given level, let the probability that the $n$th change of response type is a peak or valley at a level $l_i$ be denoted $S_i^p(n)$, $S_i^v(n)$, respectively. Then the following recurrence relations hold

$$S^p(n+1) = BS^v(n) \tag{10.15}$$

$$S^v(n+1) = AS^p(n). \tag{10.16}$$

Hence we have

$$S^p(n+2) = BAS^p(n) \tag{10.17}$$

and

$$S^v(n+2) = ABS^v(n). \tag{10.18}$$

Thus the asymptotic distributions of peaks and valleys are given by the normalized principal eigenvectors of $BA$ and $AB$.

A study of either small sample or asymptotic properties is related to a study of this finite Markov chain. This avenue of research has not been fully explored. Among some of the interesting problems yet to be solved are the following.

(i) Can some possibly approximate or conservative confidence intervals be obtained for $L_{0\cdot 50}$?

(ii) The peaks and valleys are ordered by the corresponding levels. It seems likely that an estimator based on a weighted average of peaks and valleys, with weights dependent upon the rank position of the peak (or valley), may be preferable for some purposes. (For example, if the error in $\alpha$ is large and the sequence initially goes the wrong way, an outlying peak or valley will usually result.) Such an approach would have to be made initially on the basis of asymptotic theory.

(iii) Various asymptotic properties are of interest.

The applications of this theory are of greater importance in connection with the estimation of general percentage points $L_p$.

## 10.6. Wetherill's UDTR Rule

Wetherill (1963) gave a strategy called Routine 15, for estimating general percentage points $L_p$. This is an Up-and-Down rule on a transformed response curve, and will be called the UDTR rule here.

*UDTR rule.* Use a fixed series of equally spaced levels, and after each observation, estimate the proportion $p'$ of positive responses at the level used for the current observation and consecutive to it (that is, back to the last change of response type). If $p' < p$, increase the level one step, and if $p' = p$, make no change of level. If $p' > p$, and $p'$ is based on $n_0$ trials or more, decrease the level one step.

*Ex. 10.3.* Suppose $n_0 = 3$, then the UDTR rule could be operated as follows. Move up one level after responses {0, 10 or 110} at any level. After three consecutive positive observations, move down to the next lower level. This generates a sequence of observations such as the one illustrated in Fig. 10.4.

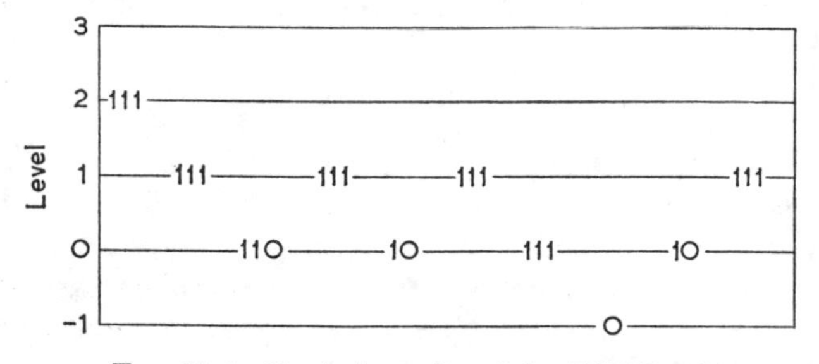


FIG. 10.4. Typical results of the UDTR rule

*Ex. 10.4.* If $n_0 = 4$, $p = 0{\cdot}75$, the UDTR rule proceeds by moving up one level after {0, 10, 110 or 11100}, and by moving down after {1111 or 11101}.

The effect of the UDTR rule is to transform the response curve. At any stage of the strategy, movement to the level above or below is determined only by a consecutive series of observations at the level, and not by the previous history of the sequence. For Ex. 10.3, at any level $x$, the probability of responses leading to a decrease in the level, is the probability of {111} which is $\{P(x)\}^3$. For Ex. 10.4, the probability of responses leading to a decrease in level is $\{P(x)\}^4\{2-P(x)\}$. In general, let the transformation be $T\{P(x)\}$. If the responses are classified merely by D or U, corresponding to responses leading to movements down or up in the level, the UDTR rule is equivalent to the Up-and-Down rule on a transformed response curve. Table 10.5 lists some possible UDTR strategies, with the corresponding transformations of the response curve.

Figure 10.5 shows how the transformations $\{P(x)\}^4$ and $\{P(x)\}^5$ would affect the standard logit response curve, and the transformed curves are for the most part similar in shape to the original curve. By

TABLE 10.5. Some possible Routine 15 strategies. Logit model†

| Entry no. | $n_0$ | Response type D | Response type U | Trans-formation | Trans. median percentage | Logit col. (5) |
|---|---|---|---|---|---|---|
| 1 | 2 | 11 | 10, 0 | $T = p^2$ | 0·7071 | 0·8812 |
| 2 | 3 | 111 | 110, 10, 0 | $T = p^3$ | 0·7940 | 1·3493 |
| 3 | 3 | 111, 1101 | 1100, 10, 0 | $T = p^3(2-p)$ | 0·7336 | 1·0131 |
| 4 | 4 | 1111 | 1110, etc. | $T = p^4$ | 0·8409 | 1·6648 |
| 5 | 5 | 11111 | 11110, etc. | $T = p^5$ | 0·8705 | 1·9054 |
| 6 | 6 | 111111 | 111110, etc. | $T = p^6$ | 0·8908 | 2·0113 |

† Extracted from Wetherill and Chen (1965).

analogy with the Up-and-Down rule, we expect the asymptotic expectation of $\overline{w}$ to be very nearly equal to the $x$ for which

$$T\{P(x)\} = 0{\cdot}50. \tag{10.19}$$

(Unless the transformed response curve is antisymmetrical about this value of $x$, the expectation of $\overline{w}$ will not be exactly equal to this value.)

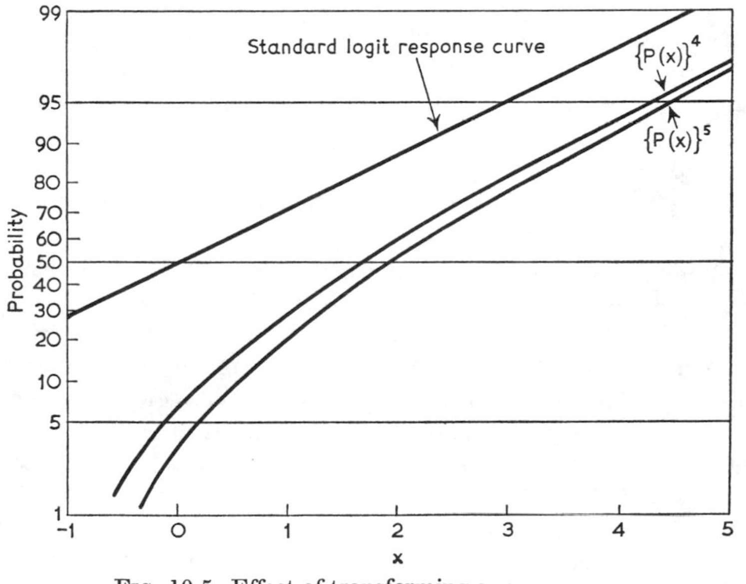


FIG. 10.5. Effect of transforming a response curve

The transformed curves of Fig. 10.5 appear to be almost linear near the transformed median percentage level (10.19). This indicates that equal spacing of levels is appropriate.

Exact calculations of $E(\overline{w})$ can be made on the basis of Markov chain theory. If $T\{P(x_j)\}$ is written instead of $p_j = P(x_j)$ in formulae (10.12) and (10.13), the theory of § 10.5 applies. Table 10.6 below lists some

TABLE 10.6 Asymptotic $E(\overline{w})$ for Routine 15†

| Table 10.5 Entry No. | Spacing | $E(\overline{w})$ | Col. 6 of Table 10.5 |
|---|---|---|---|
| 1 | 0·5 | 0·9026 | 0·8812 |
| | 1·0 | 0·9155 | |
| | 1·5 | 0·9270 | |
| 2 | 0·5 | 1·3752 | 1·3493 |
| | 1·0 | 1·3937 | |
| | 1·5 | 1·4047 | |
| 4 | 0·5 | 1·6963 | 1·6648 |
| | 1·0 | 1·7171 | |
| | 1·5 | 1·7318 | |

† Extracted from Wetherill and Chen (1965).

values of the asymptotic $E(\overline{w})$ obtained through the Markov chain theory, and they are very close to the transformed median percentage levels, except at large spacings. (In practice large spacings would not often be used for a whole sequence of observations.) Small sample $E(\overline{w})$ could be obtained in the same way, but instead, empirical sampling trials have been run comparing $\overline{w}$ estimates and maximum likelihood on a mean squared error basis. Table 10.7 below gives an extract of some of the results. For further empirical sampling trials, see Wetherill and Chen (1965). The small sample averages agree tolerably well with the transformed median percentage level. Further, on a mean squared error basis the $\overline{w}$ estimator is to be preferred to maximum likelihood estimation of the transformed median percentage level.

For final estimation from the UDTR rule, a maximum likelihood analysis could be carried out, and this method must be used if the point to be estimated is not the transformed median percentage value.

TABLE 10.7 Empirical sampling trials on Routine 15. Logit model, $\alpha = 0$, $\gamma = 1$†

| Starting value | Spacing | $w$ estimator Av. | $w$ estimator Av. (error)$^2$ | M.L. Av. (error)$^2$ | Average sample size |
|---|---|---|---|---|---|
| Table 10.5. Entry No. 1. 8 changes | | | | | |
| 0 | 0·5 | 0·7032 | 0·2008 | 0·3555 | 26·32 |
| 0 | 1·5 | 0·8774 | 0·2738 | 0·4537 | 22·04 |
| 3 | 0·5 | 1·0267 | 0·2416 | 0·3215 | 30·31 |
| 3 | 1·5 | 0·9476 | 0·2894 | 1·5164 | 23·47 |
| Table 10.5. Entry No. 4. 8 changes | | | | | |
| 0 | 0·5 | 1·4238 | 0·2133 | 0·2790 | 46·84 |
| 0 | 1·5 | 1·5964 | 0·3031 | 1·99 | 38·31 |
| 3 | 0·5 | 1·6534 | 0·1838 | 0·2145 | 49·65 |
| 3 | 1·5 | 1·6379 | 0·2802 | 1·98 | 40·07 |
| Table 10.5. Entry No. 4. 16 changes | | | | | |
| 0 | 0·5 | 1·5446 | 0·1035 | 0·1052 | 91·23 |
| 0 | 1·5 | 1·6706 | 0·1509 | 0·1750 | 74·61 |
| 3 | 0·5 | 1·6675 | 0·1053 | 0·1182 | 93·57 |
| 3 | 1·5 | 1·7054 | 0·1562 | 0·1612 | 75·73 |

† Extracted from Wetherill and Chen (1965).

However, for some industrial applications $\bar{w}$-estimation will be satisfactory.

The principal advantage of $\bar{w}$ estimation with the UDTR rule is its simplicity, combined with high efficiency. This enables a division of levels strategy to be operated, which can be defined as follows.

*Routine 17*. Classify responses as either type D or U, as given above. Proceed as in the UDTR rule until the $k$th change of response type. Then calculate the statistic $\bar{w}$ and restart the sequence near to this value, using a spacing interval one-half that used previously.

Since Routine 17 is equivalent to using Routine 11 on the transformed response curve, Routine 17 will have very similar properties to Routine 11. A suitable value for the first division of levels is about six changes. For estimation of percentage points in the tail of the response curve, a second division of levels may be worth while.

One difficulty with $\bar{w}$ estimation is that no satisfactory method of attaching a standard error to an estimate has yet been suggested. Another point for further work is that experimenters may often require some (possibly) approximate formula for the expected number of observations needed.

**Problems 10**

1. Obtain recurrence relations for $E(\overline{w})$ and $V(\overline{w})$ by using the type of argument used in deriving (10.11).

2. What characteristics of a response curve are most likely to affect $\overline{w}$ estimation?

3. As an alternative to stopping after a given number of changes of response type, Wetherill (1963) suggested stopping when the maximum number of trials at any level reached a set number. How would you compare these two stopping rules? What quantities would you study?

4. How would you associate a measure of error with $\overline{w}$ estimates?

5. Show how to calculate the asymptotic expectation and variance of the number of observations per change of response type on the UDTR rule.

# CHAPTER 11
# Double Sampling

## 11.1. Double Sampling (Sampling Inspection)

In their original paper on sampling inspection, Dodge and Romig (1929) suggested a double sampling plan, and since that date many similar sets of sampling inspection tables have included double sampling plans. (See, for example, Dodge and Romig (1944), Peach (1947), and Statistical Research Group (1948). Suppose batches of items are being inspected and that the items can be classified as good or bad, then a double sampling plan operates as follows. Take a sample of $n_1$ items and inspect, and

(a) reject the batch if the number of defectives found is greater than or equal to $c_2$;

(b) accept the batch if the number of defectives found is less than or equal to $c_1$;

(c) if neither (a) nor (b) applies, take a second sample of $n_2$ items and accept the batch if the number of defective items in the combined samples is less than or equal to $c_3$.

This sampling plan is illustrated in Fig. 11.1.

This double sampling plan requires five parameters to be set, $(n_1, n_2, c_1, c_2, c_3)$, and the choice of an optimum double sampling plan to satisfy, for example, some restrictions on the OC-curve, could be very involved. The common practice in constructing sampling inspection tables is to make some arbitrary restriction such as $n_2 = 2n_1$, or $c_2 = c_3$, etc., in order to simplify the choice. For sampling inspection use, double sampling plans could be derived through any of the varied schemes now in use, such as the 'Acceptable Quality Level' approach, 'Lot Tolerance Percent Defective', etc., see Hill (1962) for a survey of this field.

One of the most important references on double sampling is Hamaker and Van Strik (1955), in which alternative double sampling plans are compared to an 'equivalent' single sampling plan, fitted so that the 50% point of the OC-curve and the slope at this point are identical. By comparing the ASN-functions of these equivalent plans a number of important conclusions are made:

(i) The path from the origin to the critical level $c_3$ of the combined samples should pass through the gap in the first sample. (Fig. 11.1 gives a set of boundaries for which this is not true.) If this restriction is not satisfied, a more efficient plan probably exists having an 'equivalent' OC-curve.

(ii) The double sampling plans of Dodge and Romig, and many similar tables derived by using arbitrary restrictions of the type given

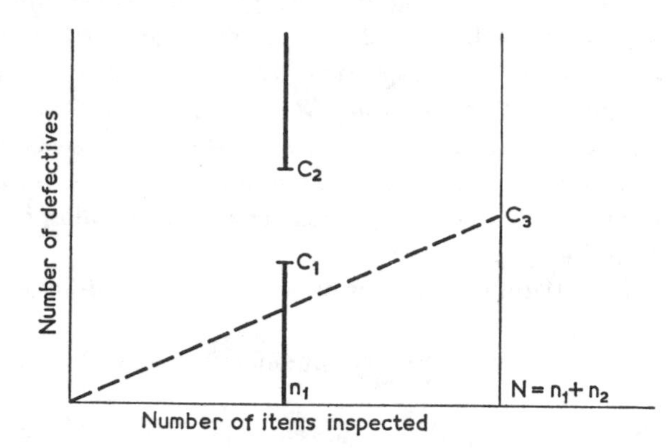


FIG. 11.1. Double sampling plan.

above, contain plans for which equivalent plans exist having a uniformly lower ASN-function.

(iii) Dodge and Romig's arbitrary restriction $c_2 = c_3$ leads to very little loss in efficiency, although Hamaker and Van Strik see no particular advantage in it, apart from the purely mathematical advantage.

Hamaker and Van Strik's recommendations were incorporated into the double sampling plans designed for the DEF-131 tables; see Hill (1962).

Hamaker and Van Strik's work should be followed up with a detailed comparison of 'equivalent' single, double, and fully sequential plans. One such comparison, but based on a decision theory framework, is given by Wetherill and Campling (1966). I conjecture that a full study should show that double sampling plans achieve most of the gain in efficiency over a single sampling which is achieved by a fully sequential plan. That is, because double sampling plans are much easier to operate than a fully sequential plan, the extra effort involved in operating a fully sequential plan may not often be worth while.

Pfanzagl (1963) gives a brief comparison of single and double sampling plans obtained by a Bayesian decision theory approach. He uses the polya prior distribution, with hypergeometric sampling. By means of calculations, he demonstrates that for a specific set of assumptions on costs, the reduction in risk obtained by using a double sampling plan is rather small.

Although this discussion has been in terms of sampling inspection, other hypothesis testing situations could be fitted into a similar framework. A double sampling plan could be designed suitable for sequential medical trials, but the fully sequential plans would probably be preferred on ethical grounds.

In this section we have assumed that, given the fact that a second sample is taken, the sample size is fixed, and not dependent on the outcome of the first sample. There is no need for this restriction, and the remainder of this chapter describes procedures in which the size of the second sample is in general a function of the outcome of the first sample.

## 11.2 Estimation by Double Sampling

Cox (1952c) described a method of designing a double sampling plan for problems involving estimation.

*Ex. 11.1.* It is required to estimate the difference $\theta$ between the mean fibre lengths of two lots of wool, the standard error of the estimate to be of the form

$$a(\theta) = k_1(1-k_2\,\mathrm{e}^{-k_3\theta^2}). \tag{11.1}$$

Suppose that the distribution of fibre length can be assumed to be normal with a known variance.

Ex. 11.1 illustrates a type of problem already discussed in § 8.1, which inevitably leads to a sequential procedure, since the number of observations desired depends on the unknown $\theta$.

A fully sequential plan could be designed by the methods of Chapter 8, but Cox is able to show that the reduction in average sample size over the methods described here must be small when large sample sizes are used (see P11.7).

If a small experiment is carried out first to provide a rough estimate of $\theta$, then the desired standard error and hence the final sample size required, can be estimated. A second experiment can now be carried out, and then $\theta$ can be estimated from the two samples. If this procedure is followed, and if the final estimate of $\theta$ is based on quantities which are unbiased estimates of $\theta$ in a fixed sample size plan, the final estimate will have a slight bias, for reasons explained in Chapter 8. Further, the standard error of the final estimate will

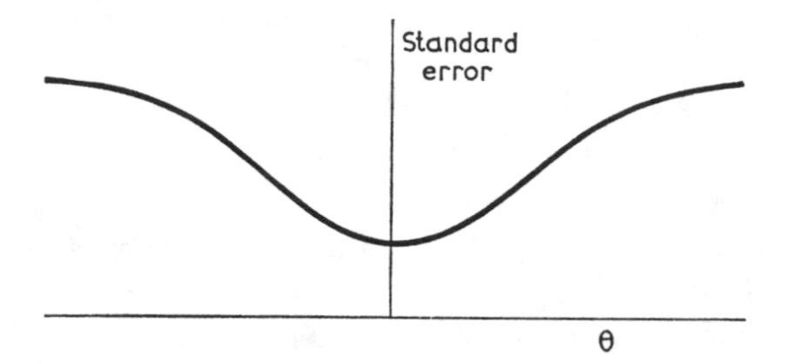


FIG. 11.2. Standard error for Ex. 11.1.

not be exactly equal to that set. Cox uses series expansion techniques to estimate a correction for bias, and to modify the final sample size slightly so as to achieve a standard error closer to that set. The difficulty with the expansion methods used, is to make sure that all terms of the same order are kept together. This is simplified by considering an infinite series of problems, as follows.

Assume that the distribution of a random variable $x$ depends on one unknown parameter $\theta$. Let $t(m)$ be an estimator of $\theta$ based on $m$ observations, and assume that it has the following properties in a fixed sample size plan,

(i) $$E\{t(m)\} = \theta$$

(ii) $$V\{t(m)\} = \sigma^2(\theta)/m, \tag{11.2}$$

where $\sigma^2(\theta)$ is a known function of $\theta$.

(iii) The skewness coefficient of $t(m)$ is asymptotically $\gamma_1(\theta)/\sqrt{m}$, where $\gamma_1(\theta)$ is known.

(iv) The coefficient of kurtosis $\gamma_2$ is asymptotically $O(m^{-1})$.

Under these conditions we shall require the final estimate of $\theta$ to have a standard error of $a(\theta)/\lambda$, where we let $\lambda$ tend to infinity in order to introduce the infinite series of problems mentioned above. In practical applications we put $\lambda = 1$.

For any fixed value of $\lambda$, let the initial sample size be $N\lambda$, and denote the estimate of $\theta$ from this sample by $t_1$. From assumption (ii), the variance of $t_1$ is

$$\sigma^2(\theta)/(N\lambda),$$

which is estimated by

$$\sigma^2(t_1)/(N\lambda) \tag{11.3}$$

Therefore the total sample size required is estimated at $\lambda n_0(t_1)$, where

$$n_0(t_1) = \sigma^2(t_1)/a(t_1).$$

In fact we shall choose instead a total sample size of $\lambda n(t_1)$, where

$$n(t_1) = n_0(t_1)\{1+b(t_1)/\lambda\}, \tag{11.4}$$

where $b(t_1)$ will be chosen so that the estimates approach more closely the desired variance. The second sample size is therefore

$$\text{Max}\{0,(n(t_1)-N)\lambda\}$$

and let the estimate of $\theta$ from the second sample only be $t_2$. Cox suggests that the combined estimate

$$t_3 = [Nt_1+\{n(t_1)-N\}t_2]/n(t_1) \tag{11.5}$$

be used to estimate $\theta$. If the quantity $t$ is a linear function of the observations, such as the sample mean, then (11.5) is just the combined estimate $t\{n(t_1)\}$, on all observations. However, if $t$ is the sample variance for example, then (11.5) is not the sample variance of the entire sample, one degree of freedom (between samples) being lost. The estimate (11.5) is used instead of $t\{n(t_1)\}$ to facilitate the mathematical development. Assume that the possibility that the first sample is already too large can be ignored, that is, we may assume $n(t_1) > N$. Then we have

$$\begin{aligned} E(t_3) &= NE\{t_1.r(t_1)\}+E\{t_2.(1-Nr(t_1))\} \\ &= NE\{t_1.r(t_1)\}+\theta-N\theta E\{r(t_1)\} \end{aligned} \tag{11.6}$$

where $r(t_1) = 1/n(t_1)$. Consider the expectations on the right-hand side of (11.6) separately, and we have

$$E\{r(t_1)\} = E\{r_0(t_1)(1-b(t_1)/\lambda)\}+O(\lambda^{-2})$$
$$= E\{r_0(t_1)\}-\lambda^{-1}E\{b(t_1)\,r_0(t_1)\}+0(\lambda^{-2}) \qquad (11.7)$$

where $r_0(t_1) = 1/n_0(t_1)$. The first term on the right-hand side of (11.7) is

$$E\{r_0(t_1)\} = E\{r_0(\theta)+(t-\theta)\,r_0'(\theta)+\tfrac{1}{2}(t-\theta)^2\,r_0''(\theta)+\ldots\}$$
$$= r_0(\theta)+\frac{\sigma^2(t_1)}{2N\lambda}r_0''(\theta)+O(\lambda^{-2}).$$

The other expectations in (11.7) and (11.6) can be evaluated in a similar way, and we obtain

$$E(t_3) = \theta+\lambda^{-1}r_0'(\theta)\,\sigma^2(\theta)+O(\lambda^{-2}).$$

Therefore an unbiased estimate of $\theta$ to the order $O(\lambda^{-1})$ is

$$t_4 = t_3-\lambda^{-1}r_0'(t_3)\,\sigma^2(t_3). \qquad (11.8)$$

By similar methods we obtain

$$V(t_4) = a(\theta).\lambda^{-1}+a(\theta).\lambda^{-2}\{K(\theta)-b(\theta)\}+O(\lambda^{-3}),$$

where

$$K(\theta) = n_0(\theta)\,\sigma^2(\theta)\{2r_0(\theta).r_0'(\theta).\gamma_1(\theta).\sigma^{-1}(\theta)+r_0'^2(\theta)$$
$$+2r_0(\theta).r_0''(\theta)+r_0''(\theta)/2N\}. \qquad (11.9)$$

If we put

$$b(\theta) = K(\theta),$$

then the estimate $t_4$ has the desired variance to the order $O(\lambda^{-2})$.

*Ex. 11.2.* Observations $x_1, x_2, \ldots, x_m$, have a Poisson distribution with mean $\theta$. It is desired to estimate $\theta$ with a variance $V(\theta) = a$.

If we use the sample mean $\bar{x}$ as an estimator of $\theta$, then we have:

(i) $$E(\bar{x}) = \theta.$$

(ii) $$V(\bar{x}) = \theta/m.$$

(iii) $$\gamma_1(\bar{x}) = 1/\sqrt{(\theta m)}.$$

The quantities $n_0(\theta)$ and $b(\theta)$ are

$$n_0(\theta) = \theta/a$$
$$b(\theta) = (3a+\theta/N)/\theta^2.$$

Let $\bar{x}_1$ be the average of the first sample, and $\bar{x}$ be the average of the combined samples, then the final sample size (11.4) is

$$n(\bar{x}_1) = \frac{\bar{x}_1}{a}\left(1+\frac{3a}{\bar{x}_1^2}+\frac{1}{\bar{x}_1 N}\right)$$

and the final estimate $t_4$ is

$$\theta = \bar{x}+a/\bar{x}. \tag{11.10}$$

This procedure breaks down for small $\theta$ since the expansions used are invalid at $\theta = 0$. The first sample size is chosen to be less than the expected value of $n(\bar{x}_1)$, but since this quantity is not known before the experiment some judgment is involved here.

The bias can be explained intuitively as follows. If, for a given value of $N$, $\bar{x}_1$ happens to be small, the second sample size will tend to be small and $\bar{x}_1$ will weigh heavily in the final estimate. On the other hand, if $\bar{x}_1$ is large, a large second sample will be taken and $\bar{x}_1$ will count less in the final estimate. Therefore the overall average $\bar{x}$ will have an expectation less than $\theta$, and the estimator (11.10) can be seen to have the right properties.

Confidence intervals and significance tests are easily incorporated into the above procedure. If the estimate $t_4$ is normally distributed, $(1-2a)\%$ confidence bounds $(\theta_l, \theta_u)$ are the solutions of

$$\begin{aligned} \theta_l + g_\alpha a^{1/2}(\theta_l) &= t_4 \\ \theta_u - g_\alpha a^{1/2}(\theta_u) &= t_4 \end{aligned} \tag{11.11}$$

where we have used $\lambda = 1$, and where

$$\int_{-\infty}^{g_\alpha} \frac{1}{\sqrt{2\pi}} \mathrm{e}^{-\frac{1}{2}u^2}\, du = 1-\alpha.$$

The null hypothesis $H_0 : \theta = \theta'$ is rejected at the $(2\alpha)\%$ level if $\theta'$ is outisde the limits $(\theta_l, \theta_u)$. In particular $H_0 : \theta = 0$ is rejected when

$$|t_4| > g_\alpha a^{1/2}(0). \tag{11.12}$$

For further examples, and a generalization to the case of estimation of a parameter $\theta$ in the presence of a nuisance parameter $\phi$, see the source paper.

To sum up, this method enables point and interval estimation and significance tests to be carried out quite simply on the same sequential test. Further, we are not restricted to one significance level as in other hypothesis testing plans given in this book. However, in practice, both the bias correction in (11.8) and the adjustment to the total sample size in (11.4) may be negligible, and the combined estimate $t_3$ may not be the usual estimate on the combined samples. Even so, the mathematics can be used to demonstrate that the double sampling plan obtained in an intuitively obvious way is approximately valid, and very efficient if the first sample size is large; see P11.7.

### 11.3. Choice of the Second Sample Size in Double Sampling

Grundy, Healy and Rees (1956) described an economic basis for choosing the second sample size in a double sampling plan, where the underlying statistical problem is a choice between two alternatives. Suppose that, for example in agriculture, a new process is suggested and a field trial is necessary to estimate the increase in output over the process in current use. Field trials are expensive, and costs could be drastically reduced by operating a double sampling plan instead of a single sampling plan. A fully sequential approach would lead to a further saving, but where experiments take some time to carry out, as in agriculture, a double sampling plan at most, would be tolerated.

Suppose the new process gives an increase in yield of $\theta$ units, and that it results in an increase in profit of $k\theta$, where $k$ is measured in monetary units, relative to the cost of an observation (which is assumed constant). If the solution is a continuing one, then $k$ can be fixed by using an equivalent capital sum at the current rate of interest. Denote the sizes of the first and second samples by $n_1$ and $n_2$, and the corresponding estimates of $\theta$ by $x_1$ and $x_2$. We shall assume that $x_1$ and $x_2$ are normally distributed with variances $\sigma^2/n_1$ and $\sigma^2/n_2$, where $\sigma^2$ is assumed known (or estimated with high precision).

After the first sample has been drawn we must decide whether a second sample is required before a decision can be made, and, if so, what size it should be. If a second sample of $n_2$ is taken (at a cost of $n_2$ cost units), the expected gain will be $Pk\theta$, where $P$ is the probability of deciding for the new process conditional on $x_1$, $n_1$ and $n_2$. Clearly this is the probability that $\{(n_1x_1+n_2x_2)/(n_1+n_2)\} > 0$ which is

$$P = \Phi\{(n_1x_1+n_2\theta)/\sigma\sqrt{n_2}\}$$

and the risk conditional on $x_1$, $n_1$, $n_2$ is

$$U(\theta|x_1, n_1, n_2) = n_2 - Pk\theta. \tag{11.13}$$

Our knowledge of $\theta$ after the first sample is taken can be represented conveniently by the posterior distribution corresponding to a uniform prior distribution at the start of the experiment. This distribution is normal with mean $x_1$ and variance $\sigma^2/n_1$, and we may now write the risk

$$U^*(x_1, n_1, n_2) = n_2 - k \int_{-\infty}^{\infty} P \frac{\beta n_1^{\frac{1}{2}}}{\sqrt{(2\pi)}\,\sigma} \exp\left\{-\frac{(\theta - x_1)^2}{2\sigma^2} n_1\right\} d\theta. \tag{11.14}$$

After some reduction this can be written

$$U^*(x_1, n_1, n_2) = n_2 - \frac{k\sigma}{\sqrt{n_1}}\left[X\Phi\{X\sqrt{(n_1+n_2)}/\sqrt{n_2}\} + \frac{\sqrt{n_2}}{\sqrt{(n_1+n_2)}} \frac{1}{\sqrt{(2\pi)}} \exp\left\{-\frac{X^2(n_1+n_2)}{2n_2}\right\}\right], \tag{11.15}$$

where $X = x_1\sqrt{n_1}/\sigma$. The use of the uniform prior distribution is reasonable here, and in any case the prior distribution used would have very little effect if $n_1$ is large enough. (Grundy *et al.* reached the same equation by using the fiducial distribution of $\theta$.)

The second sample size $n_2$ can now be chosen to make the risk a minimum. Grundy *et al.* show that the risk (11.15) either increases steadily, so that $n_2 = 0$ is the best choice, or else it increases to a (local) maximum, then decreases to a minimum before finally increasing steadily. In the latter case, the maximum and minimum are the smaller and larger roots of

$$X^2(n_2+n_1)/n_2 + \log\{n_2(n_2+n_1)^3/n_1\} = \log(k^2\sigma^2/8\pi). \tag{11.16}$$

However, it may be that the minimized risk is still larger than the risk for $n_2 = 0$. The risk at $n_2 = 0$ depends on the value of $x_1$, and is

$$U^*(x_1, n_1, 0) = \begin{cases} -kx_1 & (x_1 > 0) \\ 0 & (x_1 < 0). \end{cases}$$

Using this, Grundy *et al.* showed that the minimized risk is less than $U^*(x_1, n_1, 0)$ only if

$$\frac{|X|\sqrt{(n_1+n_2)}\,\Phi\{-|X|\sqrt{(n_1+n_2)}/\sqrt{n_2}\}\sqrt{2\pi}}{\sqrt{n_2}\exp\left\{-X^2\frac{(n_1+n_2)}{2n_2}\right\}} < \frac{2n_2+n_1}{2n_2+2n_1}. \tag{11.17}$$

The final procedure is therefore to take a second sample of size equal to the larger root of (11.16), provided that for this value of $n_2$, inequality (11.17) is satisfied.

One interesting property of this solution is that if a second sample is taken, it is of a substantial size. This is because the first sample provides some information on $\theta$, and the second sample will only be worth while if it makes a substantial contribution to the knowledge of $\theta$, sufficient to change the decision. This property is interesting in that it is not true of Cox's double sampling methods described in § 11.2.

To justify the procedure Grundy *et al.* calculated the risk function and expected sample size, and also the probability of a wrong final decision as a function of $\theta$ and the initial sample size. These properties were compared with similar properties for alternative sampling plans. Grundy *et al.* also point out the limitations of their approach. Economic considerations are assumed to be of overriding importance, and the necessary costs and profits must be known. Only a single parameter is studied, and manpower and resources are assumed not to be limiting factors.

If one treatment being compared is a standard, it would be usual to discount the value of the new process, and make an extra cost, say $kc$ for a changeover to the new treatment. This can be allowed for by simply subtracting $c$ from all observations.

## 11.4. Stein's Double Sampling Plan

No discussion of double sampling would be complete without mentioning Stein's (1945) double sampling plan for confidence intervals of a prescribed length.

Suppose random variables $x_i$ are independently and normally distributed with unknown mean and variance, $\theta$, $\sigma^2$, and the problem is to estimate $\theta$ by a $(1-\alpha)\%$ confidence interval of length less than $l$. First a sample of $n_1$ observations is taken, and we calculate the estimate $s_1^2$

$$s_1^2 = \frac{1}{n_1-1} \sum_{i=1}^{n_1} (x_i - \bar{x})^2$$

of $\sigma^2$, where $\bar{x}_1$ is the mean of this first sample. A second sample is now taken, to make the sample up to $N$ in all, and the sample mean $\bar{x}_c$ of the combined samples is calculated. Then a set of $(1-\alpha)\%$ confidence intervals can be shown to be

$$\left(\bar{x}_c - t_\alpha(n_1-1).\frac{s_1}{\sqrt{N}},\, \bar{x}_c + t_\alpha(n_1-1).\frac{s_1}{\sqrt{N}}\right) \qquad (11.18)$$

where $t_\alpha(n_1-1)$ is the two-sided $\alpha\%$ point of the $t$-distribution on $(n_1-1)$ degrees of freedom. (This confidence interval uses $s_1^2$, the sample variance of the first sample only. Another set of confidence intervals could be based on the sample variance $s_c^2$ of the combined samples.)

The length of the confidence intervals (11.18) is

$$\frac{2t_\alpha(n_1-1)\,s_1}{\sqrt{N}},$$

so that if we choose

$$N = [4t_\alpha^2(n_1-1).s_1^2/l^2]+1,$$

where $[x]$ is the greatest integer less than $x$, then the confidence intervals will have a length less than $l$. (If $N < n_1$, the confidence intervals based on the first sample alone are already less than $l$ in length.)

This method ignores information in the second sample of $(N-n_1)$ observations, on the variance $\sigma^2$. Unfortunately if we use this information, we do not necessarily have a confidence interval less than $l$ in length. In a practical application, it is unlikely that our requirement on the length of the confidence interval is so rigid as to prevent our using this additional information. This solution is typical of many in sequential analysis: by some ingenuity, a solution is obtained to a precisely stated mathematical problem, but neither the problem nor, still less, the solution corresponds to what the practising statistician really wants to do. For some similar work see Birnbaum and Healy (1960).

**Problems 11**

1. Discuss how you would set up a double sampling plan for a sequential medical trial. Start by considering a simple case, such as when observations $x_i$ are independently and normally distributed with known variance, and when it is required to do one- and two-sided tests of the null hypothesis that the mean $\theta$ is zero.

2. Observations $x_i$ are independently and normally distributed with known variance $\sigma^2$. Set up a double sampling plan to test $H_0: \theta = -\theta_1$ against $H_1: \theta = +\theta_1$, where $\theta$ is the unknown mean, and when the probabilities of error are to be $\alpha$.
(a) Obtain the double sampling plan for which the maximum average sample size is a minimum.
(b) Suppose observations cost one unit each, and wrong decisions cost an amount $K$. Obtain the double sampling plan for which the maximum expected loss is a minimum. Compare the loss function of this plan with that of the appropriate SPRT.

3. Extend Hamaker and Van Strik's work to a discussion of the efficiency of single, double and sequential sampling plans. Also discuss the relative merits of these plans in terms of loss and risk functions. In order to do the latter, you will have to derive, for example, Bayes solutions for single, double and sequential plans, based on the same underlying statistical model.

4. Independent random variables $x_i$, $i = 1, 2, \ldots$, have a normal distribution with unit variance and an unknown mean $\theta$. Obtain a double sampling procedure to estimate $\theta$ with a variance

$$V(\hat{\theta}) = a\theta^2.$$

(See Cox, 1952c, p. 220, Example 2.)

5. Independent random variables $x_i$, $i = 1, 2, \ldots$, are either one or zero, with

$$\Pr(x_i = 1) = \theta.$$

Obtain a double sampling procedure to estimate $\theta$ with variance

$$V(\hat{\theta}) = a\theta^2.$$

(See Cox, 1952c, p. 220, Example 3.)

6. Independent random variables $x_i$, $i = 1, 2, \ldots$, have a normal distribution with unit variance and unknown mean $\theta$. Obtain a double sampling procedure which satisfies the following requirements.
(i) $V(\hat{\theta}) = a(\theta)$, where $a(\theta)$ has the form (11.1).

(ii) When $\theta$ is very close to zero, $V(\hat{\theta}) \simeq 0{\cdot}015$.
(iii) When $\theta$ is very different from zero, $V(\hat{\theta}) \simeq 0{\cdot}040$.
(iv) A significance test of $H_0 : \theta = 0$ at the 5% level is to be rejected with probability 0·975 when $\theta = \pm\frac{1}{2}$.
(See Cox, 1952c, p. 225, Example 8.)

7. Show that for Cox's double sampling plan,

$$E\{\lambda n(t_1)\} = \lambda n_0(\theta) + b(\theta) . n_0(\theta) + \sigma^2(\theta) . n_0''(\theta)/2N.$$

Use the Wolfowitz generalization of the Cramer–Rao inequality (§ 8.8), to show that all sequential plans asymptotically satisfy

$$E\{\lambda n(t_1)\} \geqslant \lambda n_0(\theta).$$

Hence discuss the asymptotic efficiency of Cox's double sampling plan. (See Cox, 1952c, p. 222, § 4.)

How would you study the efficiency of the method in small samples?

8. Suppose that in Grundy *et al.*'s double sampling plan (§ 11.3), the cost of sampling in the second sample is not constant, but that the cost of each unit is one for $n_2 < N'$, and $(1+\lambda^2)$ for each unit greater than $N'$. What is the final plan in this case? (See Grundy *et al.*, 1956, p. 37.)

9. One criticism of Grundy *et al.*'s method (§ 11.3) of double sampling is that they give no guidance on how to choose the initial sample size $n_1$. Assume that the second sample is chosen by the method indicated, and show how to obtain an initial sample size which minimizes the maximum (over $\theta$) of the risk. (See Maurice in the discussion of Grundy *et al.* (1956).)

10. The double sampling plans given for sampling inspection all have the property that, given the fact that a second sample is taken, the sample size is not dependent on the outcome of the first sample. Use a decision theory formulation, similar to the one given in § 11.3, to examine the reduction in risk when the size of the second sample is a function of the outcome of the first sample.

CHAPTER 12

# Selection Procedures

## 12.1. Plant Selection

Selection procedures are a borderline subject for the present volume. Although they are necessarily sequential, it may not be desirable in practice to stick rigidly to some formalized rule. However, selection procedures are of considerable practical importance, and the first two sections of this chapter illustrate possible applications for methods described in earlier chapters. The discussion below gives only a brief outline of the subject.

Finney (1958, 1960) and Curnow (1961) have discussed plant variety selection trials, the basic outline of which is as follows. A series of new types of a species (varieties) is developed by plant breeders and these are first subjected to preliminary trials. During this phase only a small number of plants – perhaps one replicate – of each variety is used, and plants will be passed on to the next stage mainly on a basis of immunity to disease, resistance to pests, etc. (Considerations of this kind could cause rejection at any stage notwithstanding the yield.) After variety trials comes the yield selection phase, when $N$ varieties presented are reduced to $n$ on the basis of highest yield (where yield is suitably defined). Finally the selected varieties are compared with established, standard varieties, and those found markedly superior are passed into commercial use. We therefore have the following three phases,

Preliminary trials $\rightarrow$ Yield selection $\rightarrow$ Comparison with standard.

Finney (1958) suggests an alternative plan in which a standard is included at all stages, and only varieties showing a marked superiority to the standard are passed forward to the next stage. This suggestion is not followed up because, 'evaluation of its consequences seems to involve greater mathematical difficulty'. Instead Finney studies a simplified model which he admits is too rigid to apply directly in practice, but which is sufficiently realistic for the results to give general guidance on how such trials ought to be conducted.

Suppose that in the yield selection phase, $N$ varieties are to be reduced to $n$ in $k$ stages. In some cases it will be better to discard a proportion $(1-P_0)$ of the varieties without trials so that more attention can be given to the remainder. We shall assume that the discarding is random, but in practice it might be based on hunches, or on the slender evidence from the preliminary trials phase. Let the proportion selected at stage $r$ be $P_r$, so that

$$n = \pi N \tag{12.1}$$

where

$$\pi = P_0 P_1 \ldots P_k.$$

We shall assume that there is a true mean yield $x$ for each variety, and that for the $N$ varieties these are drawn randomly from a normal population with mean $\xi$ and variance $\sigma^2$. (If the initial discard is random the remaining $P_0 N$ varieties still have the same prior distribution.) Observations on yield at stage $r$ of the 'yield selection' phase will be supposed to be normally distributed with mean $x$ and variance $\epsilon_r^2$. The variances $\epsilon_r^2$ will clearly depend on the area $A_r$ available for the $r$th stage (the total area is denoted $A$) and the number of plant varieties $NP_0P_1 \ldots P_{r-1}$ remaining. Finney assumed

$$\epsilon_r^2 = \frac{P_0 P_1 \ldots P_{r-1} A \gamma \sigma^2}{A_r} \tag{12.2}$$

where $\gamma$ is a constant. This assumption makes the error variance inversely proportional to area per variety. These assumptions are all rather unrealistic; for example, selection in the preliminary trials phase may have produced a skew distribution for the true mean yield of varieties presented. However, Curnow (1961) discussed the effect of non-normality in the distribution of $x$, and suggested that unless the distribution is highly negatively skew, little is to be lost by treating it as normal in the theory.

Finney (1958) does not use the words 'prior distribution' which have been used above, nor does he use Bayesian methods. Instead the selection of varieties at any phase depends only on the results observed in that stage, and all previous results on the varieties are ignored. A

Bayesian would evaluate the posterior distribution, and through this all available information would be used. Unfortunately there are difficulties, particularly that a varieties by year interaction may exist, due to climatic differences. Curnow (1961) gives a brief discussion of the possible gain from using previous information. It seems to me that there is a point worth following up here.

Finney breaks down the problem of choosing an optimum selection plan into two stages.

(a) 'Internal economy'. For given $N$, $n$, $A$, $\gamma$ and $k$, the values of $P_0$, $P_r$, $A_r$ $(r = 1, 2, \ldots, k)$, are found which make the expectation of the means of the selected populations the greatest. The difficulty here is that in a large programme, all stages are present in any year.

(b) 'External economy'. This concerns the desirability of changes in $A$, $k$, $N$ and $n$.

For the internal economy problem with only one stage see P12.1. When two or more stages of selection are used, a mathematical difficulty arises in that once selection has taken place, the distribution of the selected populations is no longer normal, but skew, as already noted. However, Finney (1957) has been able to obtain results for the optimum $P_r$ and $A_r$, by evaluating the effect of selection on the cumulants of the distribution of yield among the varieties. These results are asymptotic in both $N$ and $n$. The external economy problem is more difficult but Finney (1960) has treated one-stage selection.

The importance of these results is that it gives a theoretical basis for a division of the total area $A$ between the stages, and for deciding the approximate proportion of varieties continued from any stage into the next. The idea is presumably that any theory is better than none, even though it is based on approximations and on a selection scheme which is too rigid to apply directly in practice.

Finney (1958) points out the error of discussing plant selection in anthropomorphic terms, such as arguing that it is 'unfair' to discard a variety not significantly poorer than those selected. He says, 'In this context, unfairness does not exist. The varieties have no personalities that require respecting, and the selectors' sole aim is to ensure that the $n$ remaining after stage $k$ are of the highest possible standard in respect of the yield criterion adopted.' This is one difference between plant selection and education selection, discussed by Finney (1962).

## 12.2. Screening for Drugs

The treatment of certain diseases in man has been revolutionized by the discovery of drugs like penicillin, and the search for new drugs for specific diseases is now sponsored widely by government and private industry both on humanitarian and financial grounds. The form the screen takes will vary with the particular disease being investigated, and the discussion below is based on screens for anti-cancer drugs. There is a large literature on the conduct of screening programmes,

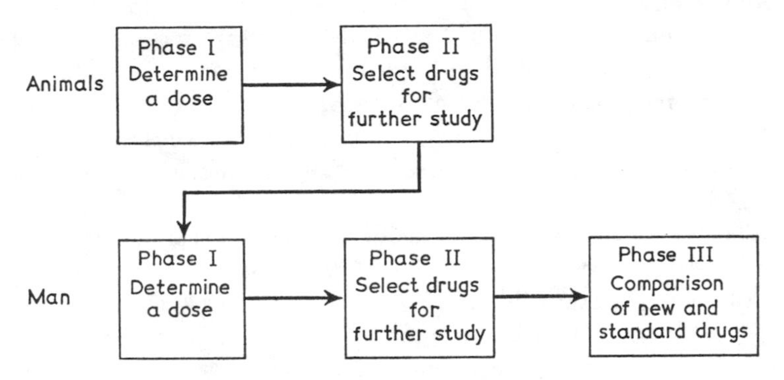


FIG. 12.1. An anti-cancer drug screen.

but I refer in particular to Armitage and Schneiderman (1958), Colton (1963a), Davies (1958, 1963), Dunnett (1960a), and Gehan (1961, 1964).

Initial trials in an anti-cancer screen must be in animals, probably rats in the first instance. The rough outline of a screen is shown in Fig. 12.1. Trials in animals can be broken up into two phases. In the first the aim is to find a dosage schedule which is not too toxic, and which can be used in further trials. In the second phase trials are conducted in order first to select drugs which are effective enough to warrant further study, and secondly to provide an estimate of their effectiveness. If these drugs are found safe in a range of animals, they are passed forward to trials in man. Trials in man can be broken up into three phases, and the first two are similar in aim to phase I and II trials in animals. First a dosage schedule which is not too toxic must be found, and then further trials made to select the drugs warranting further study. The phase III trials in man are comparative trials, to compare the effectiveness of a drug against a standard, or to select the best of a number of possible treatments. Doubtless further trials still would be made on new drugs before they are passed into general use.

The scale of these operations is very large, and a very wide range of drugs is examined, only a few of which are expected to be active. Some compounds tried may be analogues of compounds with known properties, but often the search is blind, there being no reason for examining a compound other than that it happens to be available. Gehan (1964) says that at the National Cancer Institute (U.S.A.) about 10,000 drugs are procured each year, while only approximately fifteen compounds are passed forward for further trials in man. In screening programmes of this size it is important to use efficient methods since the more efficient the trials, the greater the number of compounds that can be examined with the same facilities. The main interest statistically lies in phase II trials in animals and man, and phase III trials.

The main difference (statistically) between screening trials and plant selection lies in the length and timing of the trials. In plant selection most trials will take a season, and no more are possible until the next year. It is therefore practicable to carry forward the best varieties at each stage. In screening for drugs, trials may last only a few days, and can be repeated quickly, and it would not be convenient to wait for results on other compounds before following up a lead – a retest can be done immediately. (Only a small proportion of drugs tried have more than a negligible effect, in most cases, and no disruption of the screening programme would result from following up leads immediately.) It is therefore more convenient to operate phase II of the screening programme to carry forward the study of any drug with effectiveness greater than a specified amount, for example, effective in at least 20% of patients.

*Conduct of phase II trials*

Phase II trials in animals and man are somewhat similar, but because of ethical, political and other factors there will be differences. It will be sufficient in this brief outline to treat them together. In both cases the drugs will be passed forward to the next phase on the basis of some measure of effectiveness, without regard to relative merits of the drugs. A fully sequential trial is the most efficient in terms of observations, but since many animals can be tested at once, it is not the most efficient in time (or cost). A number of authors have suggested a two- or three-stage sampling plan, see Armitage and Schneiderman (1958), Davies (1958), Dunnett (1960a), King (1964), and Schneiderman (1961).

Assume that the measure of effectiveness of a compound is a variable $x$, which is normally distributed with unknown mean $\theta$, and known variance $\sigma^2$. In the long run, a proportion $(1-\lambda)$ of drugs will be ineffective, and have a mean close to $\theta_0$, while a proportion $\lambda$ will have effectiveness of $\theta_1$ or more. If an ineffective drug is accepted, it causes a loss equal to the cost of the further trials to which it is passed on. The rejection of effective drugs increases the average cost of development and testing of effective drugs accepted. This suggests using a Bayesian decision theory approach, with a two-point prior distribution for $\theta$, and Dunnett (1960a) obtained optimum one-, two- and three-stage plans in this way. However, a more popular approach has been to control the probabilities of accepting a drug of effectiveness $\theta_0$ and of discarding a drug of effectiveness $\theta_1$; see Dunnett and Lam (1962) and Roseberry and Gehan (1964).

*Conduct of phase III trials*

In the final stages of drug screening we might simply have a comparison of a new treatment and a standard treatment, in which case the closed sequential designs of Chapter 6 could be used. However, when we do have to choose between $k$ treatments, ethical considerations may demand a sequential trial, and we have the problem discussed in § 12.3.

Gehan (1964) gives three reasons for preferring fixed sample size plans in comparative trials. First, fixed sample size plans are easier to administer. Secondly, it frequently happens that phase III trials are not simply problems of choosing the best drug based on some measure of effectiveness, but of obtaining precise estimates of effectiveness, together with information on toxicities, etc. Thirdly, it is sometimes possible to use designs which reduce the effect of a number of different sources of variation, such as age and severity of attack. This would be difficult to do with a sequential design.

## 12.3. Selection of the Largest of $k$ Normal Means

In the third phase of drug screening trials we have the problem of choosing the best of $k$ treatments, and a similar problem arises in the third phase of plant selection trials. The problem of choosing the winner is no doubt an important one in many applications and we can formalize it as follows.

Suppose we have $k$ treatments, and we have to decide which treatment has the largest mean. Observations on the $i$th treatment are assumed to be normally distributed with mean $\mu_i$ and variance $\sigma^2$ for $i = 1, 2, \ldots, k$. The $k$ treatments are assumed to be qualitatively different, and not simply repetitions of the same treatment used in differing amounts. This latter case could be treated by the methods of Chapter 9.

In plant selection and in drug screening there will be some information available about the true configuration of the $\mu_i$ from phases I and II. In both cases too the conduct of the trials will be influenced to some extent by the fact that other treatments will become available periodically. The experiments should provide estimates and standard errors of the $\mu_i$, and be conducted in such a way that some rough comparisons with future results will be possible. In neither type of experiment will the decision depend merely on the treatment with the largest mean yield (or mean measure of effectiveness of a treatment). It will depend also on considerations of cost, side effects, etc. For phase III variety selection trials, a fixed size experiment, or at most a two-stage design, must be used, for obvious reasons. In drug screening, one of the $k$ treatments may be a standard and would be chosen unless there is good evidence that one of the other treatments is better. Thus the formalization given above will need a certain amount of modification, or some restrictions imposed, before it is applied in a particular case. However, it will serve as a basis for discussion.

The cases $\sigma^2$ known and $\sigma^2$ unknown must be dealt with separately, as a double or sequential sampling plan is required for $\sigma^2$ unknown, but not necessarily for $\sigma^2$ known. In fact a great deal of work for $\sigma^2$ known has been on fixed size experiments, but it is nevertheless of interest to study the formulations used for application in sequential procedures.

### *The indifference zone approach*

One approach, used in a series of papers by Bechhofer (see Bechhofer, 1958 and references) is to fix on the minimum probability $(1-\alpha)$ of selecting the correct population given that the difference between the largest mean and its nearest neighbour is at least a specified constant $\delta$. The actual probability of a correct selection will depend on the true configuration of treatment means, but the plans are worked out using the most unfavourable configuration, which is one mean at, say $\mu$, and $(k-1)$ together at $(\mu-\delta)$. Sampling plans can be developed for which the probability of a correct selection is at least $(1-\alpha)$ for the most unfavourable configuration. Following this approach, Bechhofer (1954) gave a single sampling plan for $\sigma^2$ known, and Bechhofer (1958) and Bechhofer and Blumenthal (1962) gave a sequential plan for $\sigma^2$ unknown. The sequential plan involves sampling all treatments until a decision is made. All these plans are based on the same approach, and prior knowledge of the configuration of means cannot be used. In a particular case, we may know from earlier trials that the most unfavourable configuration is very unlikely, but the sampling plans cannot make use of this knowledge.

*Paulson's sequential plan*

Paulson (1964) gave the following sequential procedure, which he derived using the indifference zone formulation. After each stage of sampling, take one observation from every population still included. At the $r$th stage, eliminate any population $\pi_i$ for which

$$\sum_{s=1}^{r} X_{is} < \max_{j} \left( \sum_{s=1}^{r} X_{js} \right) - a_\lambda + r\lambda,$$

where $j$ runs over all the populations still included at the $(r-1)$th stage and $X_{is}$ denotes the observation from $\pi_i$ in stage $s$. If now only one population is left, the procedure terminates; otherwise we continue sampling. Notice that once a population is eliminated, it remains eliminated, and that it is just possible for the procedure to select a population with a sample mean less than that for a population deleted at an earlier stage. The procedure is closed, since it must stop at the $(N+1)$st stage, where $N=(a_\lambda/\lambda)$, for after this, $(a_\lambda - r\lambda)$ is negative.

Paulson showed that in order to satisfy the $(\partial^*, P^*)$ requirement, we should choose

$$a_\lambda = \sigma^2/(\partial^* - \lambda) \log \{(k-1)/(1-P^*)\}$$

There is an arbitrariness in the choice of $\lambda$ in the above procedure.

Any $\lambda$ in the range $0 < \lambda < \partial^*$ will satisfy the requirement, but it is not clear what considerations, if any, one should use in choosing $\lambda$. Paulson suggests choosing $\lambda = \partial^*/4$.

An extension of Paulson's procedure to the selection of the best of $k$ Koopman–Darmois populations has been discussed by Hoel and Mazumdar (1968).

*The decision theory approach*

Dunnett (1960b) describes a Bayesian approach suitable for $\sigma^2$ known. Our knowledge of the configuration of the means $\mu_i$ at the start of phase III is represented by a set of independent normal prior distributions with means $U_i$ for $i = 1, 2, \ldots, k$ and a common (known) variance $\sigma_0^2$. Dunnett uses a fixed size experiment and assumes that a decision is made for the treatment having the largest posterior mean for $\mu_i$. Formulae are provided for

(a) the overall probability of a correct selection (integrated over the prior distribution),

(b) the conditional probability of a correct selection, given the largest mean exceeds all the others by a specified amount, and

(c) the probability that the selected mean is within a specified amount of the largest mean.

The probabilities are all multivariate normal integrals, and numerical results are only available for $k = 2$. Dunnett's approach enables an experimenter to satisfy a similar probability restriction to that in Bechhofer's method, except that prior knowledge of the configuration of means is used.

In order to extend Dunnett's results to, say, a two-stage experiment, two problems arise. First, what criterion do we use to determine the treatments included in the second stage, and secondly, how many observations are taken from each treatment in each stage, and what criterion do we use to calculate these numbers? One way to answer these questions would be via decision theory. Dunnett (1960b) gives a decision theory treatment of the choice of size for a fixed size experiment, but he does not discuss two-stage or fully sequential plans. Raiffa and Schlaifer (1961) give a very thorough discussion of the Bayesian decision theory approach to the determination of scale of effort in fixed size experiments. There is no difficulty in principle in obtaining the full sequential solution to this problem. If each population mean has a posterior mean and variance associated with it, then for $k$ populations our knowledge at any time is represented by a point in $2k$ dimensions. This space has to be divided into regions for deciding for each population, and regions for continuing sampling by selecting one observation from population $i$, for $i=1, 2, \ldots, k$. Although the theory of Chapter 7 applies to the determination of the boundaries between these regions, the amount of computing involved is very formidable, and no one has attempted it. Somerville (1954) gave a minimax approach to the determination of a fixed sample size, and also to the design of a two-stage plan for $k=3$, in which the lowest mean is (always) discarded at the first stage.

Fixed size designs are inefficient for this problem, since it will usually become clear early in the experiment that some treatments can be dropped. Maurice (1958) gave a cup-tie procedure suitable for situations in which observations must be made in pairs. A cup-tie procedure, of course, means that the treatments passed on from one stage to the next need not be those with the largest mean, since they need only be winners of a pair. It should be possible to improve considerably on Maurice's procedure when observations are not restricted to being made in pairs.

An efficient solution would probably be along the following lines. Proceed sampling until the difference between the extreme sample means is significant, then drop the smallest mean. This is repeated until only one mean is left. Unfortunately there are very great mathematical difficulties in setting up this design.

**Problems 12**

1. Finney's internal economy problem for one stage selection (see § 12.1). This problem is to choose the $P_1$, subject to fixed $N$, $n$, $A$, $\gamma$, such that the selected $n$ varieties have the largest mean. Write $n=\pi N$, where $\pi = \mathrm{P}_0\mathrm{P}_1$ and

$$P = \frac{1}{\sqrt{2\pi}} \int_{T(P)}^{\infty} \exp\{-\tfrac{1}{2}t^2\}\,\mathrm{d}t$$

$$\nu(P) = \exp\{-\tfrac{1}{2}T^2(P)\}/P\sqrt{(2\pi)}.$$

Show that the expected value of the mean of the true yields $\bar{x}_1$ for the varieties selected is

$$\bar{x}_1 = \xi + \frac{\sigma^2}{\sqrt{(\sigma^2+\epsilon_1^2)}}\,.\,\nu(P_1),$$

where $\xi$, $\sigma$ and $\epsilon_1$ are defined in § 12.1. Hence show that the optimum $P_1$ is the solution of

$$1+\frac{P_1}{P_1+\pi\gamma} = 2T(P_1)/\nu(P_1).$$

(See Finney (1958), § 9. You may assume that the mean of a fraction $P$ of variables from a normal distribution with mean $\xi$ and variance $\sigma^2$ is $\{\xi+\sigma\nu(P)\}$.)

2. Suppose two normal populations have (unknown) means $\theta_1$, $\theta_2$, and equal (known) variances $\sigma^2/2$. It is required to decide which population has the higher mean on the basis of a single sample of $n$ paired observations. Let the cost of one pair of observations be unity, and let the loss of deciding for the lowest mean be $k\theta$ where $\theta = |\theta_1-\theta_2|$. Show that the minimax choice for $n$ is

$$n = \{\tfrac{1}{2}k\sigma x\Phi(-x)\}^{2/3}$$

where

$$x\phi(x) = \Phi(-x)$$

and

$$\Phi(x) = \int_{-\infty}^{x} \phi(t)\,\mathrm{d}t$$

is the standard normal integral. (See Maurice, 1957.)

APPENDIX

# The Normal Diffusion Process

Define a process in which the increment $\Delta x$ to $x$ in a time interval $\Delta t$ has an independent normal probability density with a mean $m\Delta t$ and variance $\sigma^2 \Delta t$, denoted $\phi(\Delta x|m\Delta t, \sigma^2 \Delta t)$, where $m$ and $\sigma^2$ are constants. Write the probability that at time $t$ the value of the process lies between $x$ and $x+\Delta x$ as $f(x,t)\,\Delta x$. Then we have

$$f(y,t+\Delta t) = \int f(x,t)\,\phi(y-x|m\Delta t, \sigma^2 \Delta t)\,\mathrm{d}x. \qquad \text{(A.1)}$$

If the time intervals $\Delta t$ are unity, this is the process under study by Armitage (1957), see § 6.2. If we let $\Delta t \to 0$ then we have a continuous process known as the normal diffusion process, and this process is more amenable to mathematical treatment than the case where $\Delta t$ is discrete.

The density $\phi(y-x|m\Delta t, \sigma^2 \Delta t)$ is heavily grouped near $y-x = m\Delta t = \Delta x$ say, and if $\Delta x$ is small we have approximately

$$f(x,t) = f(y,t) - \Delta x\left(\frac{\partial f(x,t)}{\partial x}\right)_y + \frac{(\Delta x)^2}{2}\left(\frac{\partial^2 f(x,t)}{\partial x^2}\right)_y + O(\Delta x)^3. \qquad \text{(A.2)}$$

By inserting (2) into (1) we obtain

$$f(y,t+\Delta t) = f(y,t) - m\Delta t\left(\frac{\partial f(x,t)}{\partial x}\right)_y \frac{+\sigma^2 \Delta t}{2}\left(\frac{\partial^2 f(x,t)}{\partial x^2}\right)_y,$$

and by taking the limit $\Delta t \to 0$ we have

$$\frac{\partial f(x,t)}{\partial t} + m\frac{\partial f(x,t)}{\partial x} = \frac{\sigma^2}{2}\frac{\partial^2 f(x,t)}{\partial x^2} \qquad \text{(A.3)}$$

which can be shown to be an exact equation.

An unrestricted solution of (3) is clearly

$$f(x,t) = \frac{1}{\sqrt{(2\pi\sigma^2 t)}}\exp\left\{-\frac{(x-mt)^2}{2\sigma^2 t}\right\} \qquad \text{(A.4)}$$

for the initial condition that the process starts from $x = 0$ at $t = 0$. The general solution is a weighted linear combination of such solutions, starting from arbitrary points $x$.

For the application to § 6.2, it is of interest to construct the solution for a process starting at $x = 0$ when $t = 0$, with an absorbing barrier at $x = a$. The weighted linear combination needed will have zero density at $x = a$, but the drift on all components must be in the same direction, or the density for some $t$, $x < a$, would be negative.

Consider

$$f_1(x,t) = \frac{1}{\sqrt{(2\pi\sigma^2 t)}} \exp\left\{-\frac{(x-2a-mt)^2}{2\sigma^2 t}\right\}$$

then zero density along $x = -a$ is obtained by

$$g(x,t) = f(x,t) - \exp(2am/\sigma^2) f_1(x,t), \tag{A.5}$$

for $x \leqslant a$, and $g(x,t)$ is the required probability density.

The probability that the process has reached the boundary before a time $T$ is clearly

$$1 - \int_{-\infty}^{a} g(x,T)\,dx$$

or

$$1 - \Phi\left(\frac{a-mT}{\sigma\sqrt{T}}\right) + \exp(2am/\sigma^2)\,\Phi\left(\frac{-a-mT}{\sigma\sqrt{T}}\right). \tag{A.6}$$

This is the formula used by Armitage.

# References

## I. General


ALEXANDER, R., and SUICH, R. C., 1973. 'A truncated sequential $t$-test for general $\alpha$ and $\beta$'. *Technometrics*, **15**. 79–86.

AMSTER, S. J., 1963. 'A modified Bayes stopping rule.' *Ann. Math. Statist.*, **34**, 1404–1413.

ANASTASI, A., 1953. 'An empirical study of the applicability of sequential analysis to item selection.' *Educ. and Psych. Meas.*, **13**, 3–13.

ANDERSON, R. L., 1953. 'Recent advances in finding best operating conditions.' *J. Amer. Statist. Ass.*, **48**, 789–798.

ANDERSON, S. L., 1954. 'A simple method of comparing the breaking strength of two yarns.' *Journal of the Textile Inst.*, **45**, 472–479.

ANDERSON, T. W., 1960. 'A modification of the sequential probability ratio test to reduce the sample size.' *Ann. Math. Statist.*, **31**, 165–197.

ANSCOMBE, F. J., 1952. 'Large sample theory of sequential estimation.' *Proc. Camb. Phil. Soc.*, **48**, 600–607.

ANSCOMBE, F. J., 1953. 'Sequential estimation.' *Journ. Roy. Statist. Soc.*, B, **15**, 1–29.

ANSCOMBE, F. J., 1958. 'Rectifying inspection of a continuous output.' *J. Amer. Statist. Ass.*, **53**, 702–719.

ANSCOMBE, F. J., 1961. 'Rectifying inspection of lots.' *J. Amer. Statist. Ass.*, **56**, 807–823.

ANSCOMBE, F. J., 1963. 'Sequential medical trials.' *J. Amer. Statist. Ass.*, **58**, 365–383.

ARMITAGE, P., 1947. 'Some sequential tests of Student's hypothesis.' *Journ. Roy. Statist. Soc.*, Suppl., **9**, 250–263.

ARMITAGE, P., 1950. 'Sequential analysis with more than two alternative hypotheses, and its relation to discriminant function analysis.' *Journ. Roy. Statist. Soc.*, B, **12**, 137–144.

ARMITAGE, P., 1957. 'Restricted sequential procedures.' *Biometrika*, **44**, 9–26.

ARMITAGE, P., 1958. 'Numerical studies in the sequential estimation of a binomial parameter.' *Biometrika*, **45**, 1–15.

ARMITAGE, P., 1959. 'The comparison of survival rates.' *Journ. Roy. Statist. Soc.*, A, **122**, 279–292.

ARMITAGE, P., 1960. *Sequential medical trials*. Blackwell, Oxford.

ARMITAGE, P., 1963. 'Sequential medical trials: some comments on F. J. Anscombe's paper.' *J. Amer. Statist. Ass.*, **58**, 384–387.

ARMITAGE, P., 1967. Some developments in the theory and practice of sequential medical trials'. *Proc. Fifth Berkeley Symp. Math. Statist. Prob.*, **4**, 791–804.

ARMITAGE, P., MCPHERSON, C. K., and ROWE, B. C., 1969. 'Repeated significance tests on accumulating data'. *J. R. Statist. Soc., A.*, **132**, 235–244.

ARMITAGE, P., and SCHNEIDERMAN, M. A., 1958. 'Statistical problems in a mass screening program.' *Ann. N.Y. Acad. Sci.*, **76**, 896–908.

ARNOLD, K. J., 1951. (See National Bureau of Standards below.)

AROIAN, L. A., 1968. 'Sequential analysis, direct method'. *Technometrics*, **10**, 125–132.

AROIAN, L. A., and ROBISON, D. E., 1969. 'Direct methods for exact truncated sequential tests of the mean of a normal distribution'. *Technometrics*, **11**, 661–676.

BAKER, A. G., 1950. 'Properties of some tests in sequential analysis.' *Biometrika*, **37**, 334–346.

BARNARD, G. A., 1946. 'Sequential tests in industrial statistics.' *Journ. Roy. Statist. Soc.*, Suppl., **8**, 1–26.

BARNARD, G. A., 1947. 'Review of A. Wald's sequential analysis.' *J. Amer. Statist. Ass.*, **42**, 658–664.

BARNARD, G. A., 1952. 'The frequency justification of certain sequential tests.' *Biometrika*, **39**, 144–150.

BARNARD, G. A., 1954. 'Sampling inspection and statistical decisions.' *Journ. Roy. Statist. Soc.*, B, **16**, 151–174.

BARNARD, G. A., JENKINS, G. M., and WINSTEN, C. B., 1962. 'Likelihood inference and time series.' *Journ. Roy. Statist. Soc.*, A, **125**, 321–372.

BARTHOLOMEW, D. J., 1956. 'A sequential test for randomness of intervals.' *Journ. Roy. Statist. Soc.*, B, **18**, 95–103.

BARTLETT, M. S. , 1946. 'The large sample theory of sequential tests.' *Proc. Camb. Phil. Soc.*, **42**, 239–244.

BARTLETT, M. S., 1962. *Stochastic processes*. C.U.P., Cambridge.

BECHHOFER, R. E., 1954. 'A single sample multiple-decision procedure for ranking means of normal populations with known variances.' *Ann. Math. Stat.*, **25**, 16–39.

BECHHOFER, R. E., 1958. 'A sequential multiple-decision procedure for selecting the best one of several normal populations with a common unknown variance, and its use with various experimental designs.' *Biometrics*, **14**, 408–429.

BECHHOFER, R. E., and BLUMENTHAL, S., 1962. 'A sequential multiple-decision procedure for selecting the best one of several normal populations with a common unknown variance, II: Monte Carlo sampling results and new computing formulae.' *Biometrics*, **18**, 52–67.

BHATE, D. H., 1955. 'Some properties of sequential tests.' Ph.D. thesis, Univ. of London.

BICKEL, P. J., and YAHAV, J. A., 1969. 'On an A.P.O. rule in sequential estimation with quadratic loss'. *Ann. Math. Statist.*, **40**, 417–427.

BILLARD, L., and VAGHOLKAR, M. K., 1969. 'A sequential procedure for testing a null hypothesis against a two-sided alternative hypothesis'. *J. R. Statist. Soc., B.*, **31**, 285–294.

BIRNBAUM, A., 1962. 'On the foundations of statistical inference.' *J. Amer. Statist. Ass.*, **57**, 269–306.

## REFERENCES

BIRNBAUM, A., and HEALY, W. C., 1960. 'Estimates with prescribed variance based on two-stage sampling.' *Ann. Math. Statist.*, **31**, 662–676.

BLUM, J. R., 1954. 'Multidimensional stochastic approximation methods.' *Ann. Math. Statist.*, **25**, 737–744.

BOX, G. E. P., 1954. 'The exploration and exploitation of response surfaces: some general considerations and examples.' *Biometrics*, **10**, 16–60.

BOX, G. E. P., and DRAPER, N. R., 1969. *Evolutionary operation*. Wiley, New York.

BOX, G. E. P., and HUNTER, J. S., 1957. 'Multi-factor experimental designs for exploring response surfaces.' *Ann. Math. Statist.*, **28**, 195–241.

BOX, G. E. P., and WILSON, K. B., 1951. "On the experimental attainment of optimum conditions.' *Journ. Roy. Statist. Soc.*, B, **13**, 1–45.

BREAKWELL, J. V., 1956. 'Economically optimum acceptance tests.' *J. Amer. Statist. Ass.*, **51**, 243–256.

BRESLOW, N., 1969. 'On large sample sequential analysis with applications to survivorship data'. *J. Appl. Prob.*, **6**, 261–274.

BROOKS, S. H., 1958. 'A discussion of random methods of seeking maxima.' *Operations Res.*, **6**, 244–251.

BROOKS, S. H., 1959. 'A comparison of maximum seeking methods.' *Operations Res.*, **7**, 430–457.

BROSS, I., 1952. 'Sequential medical plans.' *Biometrics*, **8**, 188–205.

BROWNLEE, K. A., HODGES, J. L., and ROSENBLATT, M., 1953. 'The up-and-down method with small samples.' *J. Amer. Statist. Ass.*, **48**, 262–277.

BURGESS, G. G., 1955. 'Use of sequential analysis for determining test item difficulty level.' *Educ. and Psych. Meas.*, **15**, 80–86.

BURKHOLDER, D. L., 1956. 'On a class of stochastic approximation processes.' *Ann. Math. Statist.*, **27**, 1044–1059.

CHANG, J. H., and TUTEUR, F. B., 1971. 'A new class of adaptive array processors'. *J. Acoust. Soc. Amer.*, **49**, 639–649.

CHIEN, Y. T., and FU, K. S., 1967.' On Bayesian learning and stochastic approximation'. *I.E.E.E. Trans. Syst. Sci. Cybernetics* (U.S.A.), **SSC 3**, 28–38.

CHOI, S. C., 1971. 'Sequential test for correlation coefficients'. *J. Amer. Statist. Ass.*, **66**, 575–576.

CHUNG, K. L., 1954. 'On a stochastic approximation method.' *Ann. Math. Statist.*, **25**, 463–483.

COLTON, T., 1963a. 'Optimal drug screening plans.' *Biometrika*, **50**, 31–46.

COLTON, T., 1963b. 'A model for selecting one of two medical treatments.' *J. Amer. Statist. Ass.*, **58**, 388–400.

COOKE, P. J., 1971. 'Sequential estimation in the uniform density'. *J. Amer. Statist. Ass.*, **66**, 614–617.

CORNELIUSSEN, A. H., and LADD, D. W., 1970. 'On sequential tests of the binomial distribution', *Technometrics* **12**, 635–646.

COWDEN, D. J., 1946. 'An application of sequential sampling to testing students.' *J. Amer. Statist. Ass.*, **41**, 547–556.

COX, C. P., and ROSEBERRY, T. D., 1966a. 'A note on the variance of the distribution of sample number in sequential probability ratio tests'. *Technometrics*, **8**, 700–704.

COX, C. P., and ROSEBERRY, T. D., 1966b. 'A large sample sequential test, using concomitant information, for discrimination between two composite hypotheses'. *J. Amer. Statist. Ass.*, **61**, 357–367.

COX, D. R., 1952a. 'Sequential tests for composite hypotheses.' *Proc. Camb. Phil. Soc.*, **48**, 290–299.

COX, D. R., 1952b. 'A note on the sequential estimation of means.' *Proc. Camb. Phil. Soc.*, **48**, 447–450.

COX, D. R., 1952c. 'Estimation by double sampling.' *Biometrika*, **39**, 217–227.

COX, D. R., 1955. 'Some statistical methods connected with series of events.' *Journ. Roy. Statist. Soc.*, B, **17**, 129–164.

COX, D. R., 1958. *Planning of experiments*. John Wiley, New York.

COX, D. R., 1963. 'Large sample sequential tests for composite hypotheses.' *Sankhya*, **25**, 5–12.

CRAMER, H., 1946. *Mathematical methods of statistics*. Princeton.

CURNOW, R. N., 1961. 'Optimal programmes for varietal selection.' *Journ. Roy. Statist. Soc.*, B, **23**, 282–318.

DAVID, H. T., and KRUSKAL, W. H., 1956. 'The WAGR sequential *t*-test reaches a decision with probability one.' *Ann. Math. Statist.*, **27**, 797–805.

DAVIES, O. L., 1956. *The design and analysis of industrial experiments*. Oliver and Boyd, Edinburgh.

DAVIES, O. L., 1958. 'The design of screening tests in the pharmaceutical industry.' *Bull. Int. Statist. Inst.*, **36**, III, 226–241.

DAVIES, O. L., 1963. 'The design of screening tests.' *Technometrics*, **5**, 483–489.

DAVIS, M., 1963. 'Comparison of sequential experiments for estimating the dosage-response curve.' Abstract 890, *Biometrics*, **19**, 504.

DAVIS, M., 1971. 'Comparison of sequential bioassays in small samples'. *J. R. Statist. Soc., B.*, **33**, 78–87.

DEGROOT, M. H., 1962. 'Uncertainty, information and sequential experiments.' *Ann. Math. Statist.*, **33**, 404–419.

DEGROOT, M. H., and NADLER, J., 1958. 'Some aspects of the use of the sequential probability ratio test.' *J. Amer. Statist. Ass.*, **53**, 187–199.

DEGROOT, M. H., and RAO, M. M., 1963. 'Bayes estimation with convex loss.' *Ann. Math. Statist.*, **34**, 839–846.

DERMAN, C., 1956. 'Stochastic approximation.' *Ann. Math. Statist.*, **27**, 879–886.

DERMAN, C., and SACHS, J., 1959. 'On Dvoretzky's stochastic approximation theorem.' *Ann. Math. Statist.*, **30**, 601–606.

DIXON, W. J., and MOOD, A. M., 1948. 'A method for obtaining and analyzing sensitivity data.' *J. Amer. Statist. Ass.*, **43**, 109–126.

DODGE, H. F., and ROMIG, H. G., 1929. 'A method of sampling inspection.' *Bell Syst. Tech. J.*, **8**, 613–631.

DODGE, H. F., and ROMIG, H. G., 1944. *Sampling inspection tables – single and double sampling*. John Wiley, New York.

DUNCAN, A. J., 1959. *Quality control and industrial statistics*. Revised ed., Irwin, Illinois.

DUNNETT, C. W., 1960a. 'Statistical theory of drug screening' in *Quantitative methods in Pharmacology*, pp. 212–231. H. de Jonge (ed.), North Holland, Amsterdam

DUNNETT, C. W., 1960b. 'On selecting the largest of $k$ normal population means.' *Journ. Roy. Statist. Soc.*, B, **22**, 1–40.

DUNNETT, C. W., and LAM, R. A., 1962. 'Sequential procedures for drug screening.' Presented at annual meeting A.S.A., Minneapolis.

DVORETZKY, A., 1956. 'On stochastic approximation.' *Proc. Third Berkeley Symp.*, **1**, 39–55, Univ. of California, Berkeley.

EL-SAYYAD, G. M., and FREEMAN, P. R., 1973. 'Bayesian sequential estimation of a Poisson process rate.' *Biometrika*, **60**, 289–296.

ENRICK, N. L., 1946. 'The U.S. Army Quartermaster Corps' use of sequential sampling inspection.' *Industrial Quality Control*, **2** (5), 12–14.

EPSTEIN, B., and SOBEL, M., 1955. 'Sequential life tests in the exponential case.' *Ann. Math. Statist.*, **26**, 82–93.

FABIAN, V., 1967. 'Stochastic approximation of minima with improved speed.' *Ann. Math. Statist.*, **38**, 191–200.

FABIAN, V., 1969. 'Stochastic approximation for smooth functions.' *Ann. Math. Statist.*, **40**, 299–302.

FEDERER, W. T., and BALAAM, L. N., 1972. *Bibliography on experiment and treatment design pre-1968*. Oliver and Boyd, Edinburgh.

FINNEY, D. J., 1957. 'The consequences of selection for a variate subject to errors of measurement.' *Reveu de l'Institut International de Statistique*, **24**, 22–29.

FINNEY, D. J., 1958. 'Statistical problems in plant selection.' *Bull. Int. Statist. Inst.*, **36**, III, 242–268.

FINNEY, D. J., 1960. 'A simple example of the external economy of varietal selection.' *Bull. Int. Statist. Inst.*, **37**, III, 91–106.

FINNEY, D. J., 1962. 'The statistical evaluation of educational allocation and selection.' *Journ. Roy. Statist. Soc.*, A, **125**, 525–564.

FISHER, R. A., 1956. *Statistical methods and scientific inference*. Oliver and Boyd, London.

FRASER, D. A. S., 1957. *Nonparametric methods in statistics*. John Wiley, New York.

FREEMAN, P. R., 1970. 'Optimal Bayesian sequential estimation of the median effective dose'. *Biometrika*, **57**, 79–90.

FREEMAN, P. R., 1972. 'Sequential estimation of the size of a population'. *Biometrika* **59**, 9–18.

FRIEDMAN, M., and SAVAGE, L. J., 1947. 'Planning experiments seeking maxima.' Chapter 13 of *Techniques of statistical analysis*. Eisenhart, C., Hastay, and Wallis (eds.), McGraw Hill, New York.

FU, K. S., 1967. 'On learning techniques in engineering cybernetic systems'. *Cybernetica* **10**, 194–213.

GEHAN, E. A., 1961. 'The determination of the number of patients required in a preliminary and a follow-up trial of a new chemotherapeutic agent.' *Journal of Chronic Diseases*, **13**, 346–353.

GEHAN, E. A., 1964. 'Clinical trials in cancer research.' Unpublished manuscript.

GHOSH, B. K., 1967. 'Sequential analysis of variance under random and mixed models'. *J. Amer. Statist. Ass.*, **62**, 1401–1417.

GHOSH, B. K., 1970. *Sequential tests of statistical hypotheses*. Addison-Wesley, Reading Mass.

GHOSH, J., 1960. 'On some properties of sequential *t*-tests.' *Bull. Calcutta Statist. Ass.*, **9**, 77–86.

GHOSH, J., 1962. 'On the monotonicity of the OC of a class of sequential probability ratio tests.' *Bull. Calcutta Statist. Ass.*, **9**, 139–144.

GILCHRIST, W. G., 1961. 'Some sequential tests using range.' *Journ. Roy. Statist. Soc.*, B, **23**, 335.

GIRSHICK, M. A., MOSTELLER, F., and SAVAGE, L. J., 1946. 'Unbiased estimates for certain binomial sampling schemes with applications.' *Ann. Math. Statist.*, **17**, 13–23.

GRIFFITHS, S., and RAO, A. G., 1965. 'An application of least cost acceptance sampling schemes.' Unternehmensforschung, (Operations Research) **9**, (1), 8-17.

GRUNDY, P. M., HEALY, M. J. R., and REES, D. H., 1956. 'Economic choice of the amount of experimentation.' *Journ. Roy. Statist. Soc.*, B, **18**, 32–55.

GUTTMAN, L., and GUTTMAN, R., 1959. 'An illustration of the use of stochastic approximation.' *Biometrics*, **15**, 551–559.

HAJNAL, J., 1960. 'Sequential trials of analgesics in rheumatoid arthritis' in *Quantitative methods in Pharmacology*. H. de Jonge (ed.), North-Holland, Amsterdam.

HAJNAL, J., 1961. 'A two-sample sequential *t*-test.' *Biometrika*, **48**, 65–75.

HAJNAL, J., SHARP, J., and POPERT, A. J., 1959. 'A method for testing analgesics in rheumatoid arthritis using a sequential procedure.' *Ann. Rheum. Dis.*, **18**, 189–206.

HALD, A., 1952. *Statistical theory with engineering applications*. John Wiley, New York.

HALD, A., 1960. 'The compound hypergeometric distribution and a system of single sampling inspection plans based on prior distributions and costs.' *Technometrics*, **2**, 275–340.

HALDANE, J. B. S., 1945. 'A labour-saving method of sampling.' *Nature, Lond.*, **155**, 49–50.

HALL, W. J., WIJSMAN, R. A., and GHOSH, B. K., 1965. 'The relationship between sufficiency and invariance with applications in sequential analysis'. *Ann. Math. Statist.*, **36**, 575–614.

HAMAKER, H. C., and VAN STRIK, R., 1955. 'The efficiency of double sampling for attributes.' *J. Amer. Statist. Ass.*, **50**, 830–849.

HECHT, B., 1947. 'Process control methods.' *Industrial Quality Control*, **4**, 7–11.

HERZBERG, A. M., and COX, D. R., 1969. 'Recent work on the design ef experiments: A bibliography and a review'. *J. R. Statist. Soc., A.*, **132**, 29–67.

HILL, I. D., 1960. 'The economic incentive provided by inspection.' *Appl. Statist.*, **9**, 69–81.

HILL, I. D., 1962. 'Sampling inspection and defence specification DEF-131.' *Journ. Roy. Statist. Soc.*, A, **125**, 31–87.

HODGES, J. L., JR., and LEHMANN, E. L., 1956. 'Two approximations to the Robbins–Monro process.' *Proc. Third Berkeley Symp.*, **1**, 95–104, Univ. of California, Berkeley.

HOEL, P. G., 1954. 'On a property of the sequential *t*-test.' *Skandinavisk Aktuarietidskrift*, **37**, 19–22.

HOEL, D. G., and MAZUMDAR, M., 1968. 'An extension of Paulson's selection procedure.' *Ann. Math. Statist.*, **39**, 2069–2074.

HORSNELL, G., 1957. 'Economical acceptance sampling schemes.' *Journ. Roy. Statist. Soc.*, A, **120**, 148–201.

IFRAM, A., 1965. 'On the asymptotic behaviour of densities with applications to sequential analysis'. *Ann. Math. Statist.*, **36**, 615–637.

JACKSON, J. E., and BRADLEY, R. A., 1961a. 'Sequential $\chi^2$ and $T^2$-tests.' *Ann. Math. Statist.*, **32**, 1063–1077.

JACKSON, J. E., and BRADLEY, R. A., 1961b. 'Sequential $\chi^2$ and $T^2$-tests and their application to an acceptance sampling problem.' *Technometrics*. **3**. 519–534.

JOANES, D. N., 1972. 'Sequential tests of composite hypotheses' *Biometrika* **59**, 633–638; *Biometrika* **62**, 1975.

JOHNSON, N. L., 1953. 'Some notes on the application of sequential methods in the analysis of variance.' *Ann. Math. Statist.*, **24**, 614–623.

JOHNSON, N. L., 1960. 'On the choice of a sequential procedure.' *Proc. Biometric Society Symp.*, pp. 27–40, Leiden.

JOHNSON, N. L., 1961. 'Sequential analysis: a survey.' *Journ. Roy. Statist. Soc.*, A, **124**, 372–411.

JONES, H. L., 1947. 'Sampling plans for verifying clerical work.' *Industrial Quality Control*, **3** (4), 5–11.

KEMP, K. W., 1958. 'Formulae for calculating the operating characteristic and the average sample number of some sequential tests.' *Journ. Roy. Statist. Soc.*, B, **20**, 379–386.

KENDALL, M. G., and STUART, A., 1961. *The advanced theory of statistics*. Vol. 2, Griffin, London.

KESTEN, H., 1958. 'Accelerated stochastic approximation.' *Ann. Math. Statist.*, **29**, 41–59.

KIEFER, J., and WOLFOWITZ, J., 1952. 'Stochastic estimation of the maximum of a regression function.' *Ann. Math. Statist.*, **23**, 462–466.

KILPATRICK, G. S., and OLDHAM, P. D., 1954. 'Calcium chloride and adrenaline as bronchial dilators compared by sequential analysis.' *Brit. Med. J.*, **ii**, 1388–1391.

KIMBALL, A. W., 1950. 'Sequential sampling plans for use in psychological test work.' *Psychometrika*, **15**, 1–15.

KING, E. P., 1964. 'Optimal replication in sequential drug screening.' *Biometrika*, **51**, 1–10.

KNIGHT, W., 1965. 'A method of sequential estimation applicable to the hypergeometric, binomial, Poisson and exponential distributions.' *Ann. Math. Statist.*, **36**, 1494–1503.

KOWALSKI, C. J., 1971. 'The OC and ASN functions of some SPRT's for the correlation coefficient'. *Technometrics*, **13**, 833–841.

LEVITT, H., 1964. 'Discrimination of sounds in hearing.' Ph.D. thesis, Univ. of London.

LINDLEY, D. V., and BARNETT, B. N., 1965. 'Sequential sampling: two decision problems with linear losses for binomial and normal variables'. *Biometrika*, **52**, 507–532.

LINNIK, YU, V., and ROMANOVSKY, I. V., 1972. 'Some new results in sequential estimation theory'. *Proc. Sixth Berk. Symp. Math. Statist. Prob.*, Vol. 1, Univ. of California Press, Berkeley.

MANTY, B. F. J., 1969. 'Approximations to the characteristics of some sequential tests'. *Biometrika*, **56**, 203–206.

MAURICE, R. J., 1957. 'A minimax procedure for choosing between two populations using sequential sampling.' *Journ. Roy. Statist. Soc.*, B, **19**, 255–261.

MAURICE, R. J., 1958. 'Selection of the population with the largest mean when comparisons can be made only in pairs.' *Biometrika*, **45**, 581–586.

MAURICE, R. J., 1959. 'A different loss function for the choice between two populations.' *Journ. Roy. Statist. Soc.*, B, **21**, 203–213.

MCPHERSON, C. K., and ARMITAGE, P., 1971. 'Repeated significance tests on accumulating data when the null hypothesis is not true'. *J. R. Statist. Soc., A.*, **134**, 15–25.

MILLER, H. D., 1961. 'A generalisation of Wald's identity with applications to random walks.' *Ann. Math. Statist.*, **32**, 549–560.

MORGAN, M. E., MACLEOD, P., ANDERSON, E. O., and BLISS, C. I., 1951. *A sequential procedure for grading milk by microscopic counts.* Conn. (Storrs) Agric. Exp. Sta. Bull., **276**, 35 pp.

MORRIS, K. W., 1963. 'A note on direct and inverse sampling.' *Biometrika*, **50**, 544–545.

NATIONAL BUREAU OF STANDARDS, 1951. *Tables to facilitate sequential* t-*tests.* Applied Math. Series No. 7, prepared by K. J. Arnold.

NEYMAN, J., and PEARSON, E. S., 1928. 'On the use and interpretation of certain test criteria for purposes of statistical inference.' *Biometrika*, **20A**, Part I, 175–240; Part II, 263–294.

OAKLAND, G. B., 1950. 'An application of sequential analysis to whitefish sampling.' *Biometrics*, **6**, 59–67.

## REFERENCES


O'BRIEN, P. C., 1973. 'A procedure for sequential interval estimation of the shape parameter of the gamma distribution'. *Technometrics*, **15**, 563–570.

PAGE, E. S., 1954. 'An improvement to Wald's approximation for some properties of sequential tests.' *Journ. Roy. Statist. Soc.*, B, **16**, 136–139.

PATNAIK, P. B., 1949. 'The non-central $\chi^2$ and $F$-distributions and their applications.' *Biometrika*, **36**, 202–232.

PAULSON, E., 1969. 'Sequential interval estimation for the means of normal populations'. *Ann. Math. Statist.*, **40**, 509–516.

PEACH, P., 1947. *An introduction to statistics and quality control.* 2nd Ed. Edwards and Broughton, Raleigh, N.C.

PFANZAGL, J., 1963. 'Sampling procedures based on prior distributions and costs.' *Technometrics*, **5**, 47–62.

PHATARFOD, R. N., 1971. 'A sequential test for gamma distributions'. *J. Amer. Statist. Ass.*, **66**, 876–878.

RAGHAVACHARI, M., 1965. 'Operating characteristic and expected sample size of a sequential probability ratio test for the simple exponential distribution.' *Bull. Calcutta Statist. Ass.*, **14**, 65–73.

RAIFFA, H., and SCHLAIFER, R., 1961. *Applied statistical decision theory.* Boston, Graduate School of Business Administration, Harvard Univ.

RAY, S. N., 1965. 'Bounds on the maximum sample size of a Bayes sequential procedure'. *Ann. Math. Statist.*, **36**, 859–878.

RAY, W. D., 1956. 'Sequential analysis applied to certain experimental designs in the analysis of variance'. *Biometrika*, **43**, 388–403.

RAY, W. D., 1957. 'Sequential confidence intervals for the mean of a normal population with unknown variance.' *Journ. Roy. Statist. Soc.*, B, **19**, 133–143.

RIVETT, B. H. P., 1951. 'Sequential analysis of machine performance' in *Statistical Method in Industrial Production*, pp. 86–89. London, Royal Statistical Society.

ROBBINS, H., 1955. 'An empirical Bayes approach to statistics.' *Proc. Third Berkeley Symp. on Statistics and Prob.*, **1**, 157–164.

ROBBINS, H., 1964. 'The empirical Bayes approach to statistical decision problems.' *Ann. Math. Statist.*, **35**, 1–20.

ROBBINS, H., and MONRO, S., 1951. 'A stochastic approximation method.' *Ann. Math. Statist.*, **22**, 400–407.

ROSEBERRY, T. D., and GEHAN, E. A., 1964. 'Operating characteristic curves and accept–reject rules for two and three stage screening procedures.' *Biometrics*, **20**, 73–84.

RUSHTON, S., 1950. 'On a sequential $t$-test.' *Biometrika*, **37**, 326–333.

RUSHTON, S., 1952. 'On a two-sided sequential $t$-test.' *Biometrika*, **39**, 302–308.

SACHS, J., 1958. 'Asymptotic distribution of stochastic approximation procedures.' *Ann. Math. Statist.*, **29**, 373–405.

SACHS, J., 1963. 'Generalised Bayes solutions in estimation problems.' *Ann. Math. Statist.*, **34**, 751–769.

SAINSBURY, P., and LUCAS, C. J., 1959. 'Sequential methods applied to the study of prochlorperazine.' *Brit. Med. J.*, **ii**, 737–740.

SCHAFER, R. E., and TAKENAGA, R., 1972. 'Sequential probability ratio test for availability.' *Technometrics*, **14**, 123–135.

SCHNEIDERMAN, M. A., 1961. 'Statistical problems in the screening search for anti-cancer drugs by the National Cancer Institute of the United States' in *Quantitative methods in Pharmacology*. H. de Jonge (ed.), North-Holland, Amsterdam.

SCHNEIDERMAN, M. A., and ARMITAGE, P., 1962a. 'A family of closed sequential procedures.' *Biometrika*, **49**, 41–56.

SCHNEIDERMAN, M. A., and ARMITAGE, P., 1962b. 'Closed sequential *t*-tests.' *Biometrika*, **49**, 359–366.

SISKIND, V., 1964. 'On certain formulae applied to the sequential *t*-test.' *Biometrika*, **51**, 97–106.

SNOKE, L. R., 1956. 'Specific studies on soil-block procedure for bioassay of wood preservatives.' *Applied Microbiology*, **4**, 21–31.

SOBEL, M., and WALD, A., 1949. 'A sequential decision procedure for choosing one of three hypotheses concerning the unknown mean of a normal distribution.' *Ann. Math. Statist.*, **20**, 502–522.

SOMERVILLE, P. N., 1954. 'Some problems of optimum sampling.' *Biometrika*, **41**, 420–429.

STATISTICAL RESEARCH GROUP, 1948. *Sampling Inspection*. McGraw-Hill, New York.

STEIN, C., 1945. 'A two-sample test for a linear hypothesis whose power is independent of the variance.' *Ann. Math. Statist.*, **16**, 243–258.

SUICH, R., and IGLEWICZ, B., 1970. 'A truncated sequential *t*-test'. *Technometrics*, **12**, 789–798.

TOCHER, K.D., 1963. The art of simulation. E.U.P., London

TWEEDIE, M. C. K., 1945. 'Inverse statistical variates.' *Nature, Lond.*, **155**, 453.

TWEEDIE, M. C. K., 1957. 'Statistical properties of inverse Gaussian distributions, I.' *Ann. Math. Statist.*, **28**, 362–377.

VAGHOLKAR, M. K., and WETHERILL, G. B., 1960. 'The most economical binomial sequential probability ratio test.' *Biometrika*, **47**, 103–109.

VENTNER, J. H., 1966. 'On Dvoretzky stochastic approximation theorems.' *Ann. Math. Statist.* **37**, 1534–1544.

WALD, A., 1947. *Sequential analysis*. John Wiley, New York.

WALD, A., 1950. *Statistical decision functions*. John Wiley, New York.

WALD, A., and WOLFOWITZ, J., 1948. 'Optimum character of the sequential probability ratio test.' *Ann. Math. Statist.*, **19**, 326–339.

WATERS, W. E., 1955. 'Sequential analysis of forest insect surveys.' *Forest Science*, **1**, 68–79.

WETHERILL, G.B., 1959. 'The most economical sequential sampling scheme for inspection by variables.' *Journ. Roy. Statist. Soc.*, B., **21**, 400-408.

WETHERILL, G. B., 1960. 'Some remarks on the Bayesian solution of the single sample inspection scheme.' *Technometrics*, **2**, 341–352.

WETHERILL, G. B., 1961. 'Bayesian sequential analysis.' *Biometrika*, **48**, 281–292.

WETHERILL, G. B., 1961. 'Sequential estimation of quantal response curves.' *Journ. Roy. Statist. Soc.*, B., **25**, 1–48.

WETHERILL, G. B., and CAMPLING, G. E. G., 1966. 'The decision theory approach to sampling inspection'. *J. R. Statist. Soc. B.*, **28**, 381–416.

WETHERILL, G. B., and CHEN, H., 1965. 'Sequential estimation of quantal response curves. II. A new method of estimation.' Tech. Memo 65–1215–1, Bell Tel. Labs., N.J.

WHITTLE, P., 1954. 'Optimum preventive sampling.' *J. Operat. Res. Soc. Amer.*, **2**, 197.

WIJSMAN, R. A. General proof of termination with probability one of invariant sequential probability ratio tests based on multivariate normal observations. *Ann. Math. Statist.*, **38**, 8–24.

WOLFOWITZ, J., 1947. 'The efficiency of sequential estimates and Wald's equation for sequential processes.' *Ann. Math. Statist.*, **18**, 215–230.

WOLFOWITZ, J., 1956. 'On stochastic approximation methods.' *Ann. Math. Statist.*, **27**, 1151–1156.

WOLFOWITZ, J., 1966. Remark on the optimum character of the sequential probability ratio test. *Ann. Math. Statist.*, **37**, 726–727.

## II. Applications

(References not given in detail are listed above.)

*Chapter 2 – Expository*

BRADLEY, R. A., 1953. 'Some statistical methods in taste test evaluation.' *Biometrics*, **9**, 22–28.

BURR, I. W., 1954. 'Fundamental principles of sequential analysis.' *Industrial Quality Control*, **9**, (6) 92–98.

BURR, I. W., 1949. 'A new method for approving a machine or process setting.' *Industrial Quality Control*, Part 1, **5** (4), 12–18; Part 2, **6** (2), 15–19; Part 3, **6** (3), 13–17.

DAVIES, O. L., 1958. 'The design of screening tests in the pharmaceutical industry.' *Bull. Int. Statist. Inst.*, **36**, III, 226–240.

HARRISON, S., and ELDER, L. W., 1950. 'Some applications of statistics to laboratory taste testing.' *Food Technology*, **4**, 434–439.

LIEBERMAN, A., 1959. 'Sequential life testing plans for the exponential distribution.' *Industrial Quality Control*, **16** (1), 14–18.

MOSHMAN, J., 1958. 'The application of sequential estimation to computer simulation and Monte Carlo procedures.' *Journ. Ass. Comp. Math.*, **5**, 343–352.

NOEL, R. H., 1952. 'Sampling for the drug industry.' *Drug and Allied Industries*, **38** (May 1952), 15.

*Other applications*

COWDEN (1946); ENRICK (1946); JONES (1947); KIMBALL (1950);

MORGAN *et al.* (1951); OAKLAND (1950); RIVETT (1951).

*Chapter 3*

ANDERSON (1954); BURGESS (1955); WATERS (1955).

*Chapter 5 – Sequential* t*-tests*

HAJNAL (1960); HAJNAL *et al.* (1959); KILPATRICK and OLDHAM (1954).

*Chapter 5 – Sequential* $\chi^2$ *and* $T^2$ *tests*

JACKSON and BRADLEY (1961b).

*Chapter 6 – Medical applications*†

BROWN, A., *et al.*, 1960. *Lancet*, **ii**, 227–230.
CALNAN, J., and BARR, A., 1960. *Brit. Med. J.*, **ii**, 261–263.
COBBAN, K. MCL., COLLARD, P. J., *et al.*, 1963. *Brit. Med. J.*, **i**, 794–796.
FLAVELL MATTS, S. G., 1960. *Lancet*, **i**, 517–519.
FREIREICH, E. J., *et al.*, 1963. *Blood*, **21**, 699–716.
HELLIER, F. F., 1963. *Lancet*, **i**, 471–472.
MARHSALL, J., and SHAW, D. A., 1960. *Lancet*, **i**, 995–998.
MURPHY, F. M., BARBER, J. M., and KILPATRICK, S. J., 1961. *Brit. Med. J.*, **i**, 139–140.
NEWTON, D. R. L., and TANNER, J. M., 1956. *Brit. Med. J.*, **ii**, 1096–1099. (Bross plan B.)
PARSONS, T. W., and THOMPSON, T. J., 1961. *Brit. Med. J.*, **i**, 171–173. (Bross plan B.)
ROBERTSON, J. D., and ARMITAGE, P., 1959. *Anaesthesia*, **14**, 53–64.
RUSSELL, B., FRAIN-BELL, W., *et al.*, 1960. *Lancet*, **i**, 1141–1147.
SNELL, E. S., and ARMITAGE, P., 1957. *Lancet*, **i**, 860–862.
THOMPSON, T. J., 1958. *Brit. Med. J.*, **ii**, 1140–1141. (Bross plan B.)
TRUELOVE, S. C., WATKINSON, G., and DRAPER, G., 1962. *Brit. Med. J.*, **ii**, 1708–1711.
VAKIL, B. J., TULPULE, T. H., *et al.*, 1963. *Clin. Pharmacol. Therapeut.*, **4**, 182–187.
WHITTLE, C. H., WOODS, B., *et al.*, 1961. *Brit. J. Dermatol.*, **73**, 433–438.

† Except where stated, the trials were restricted Armitage binomial sampling plans.

*Chabter 9 – Robbins–Monro*

GUTTMAN and GUTTMAN (1959).
HAWKINS, D. F., 1964. 'Observations on the application of the Robbins–Monro process to sequential toxicity assays.' *Br. J. Pharmac. Chemother.*, **22**, 392–402.
LORD, F. M., 1971. 'Robbins–Monro procedures for tailored testing'. *Educ. Psych. Measurement*, **31**, 3–31.

## REFERENCES

*Chapter 10 – Up-and-Down rule*

LORD, F. M., 1971. 'Tailored testing, application of stochastic approximation'. *J. Amer. Statist. Ass.*, **66**, 707–711.

RÜMKE, C. L., 1959. 'The influence of atropine on the toxicity of thialbarbitone, thiopentone and pentobarbital.' *Arch. int. Pharmacodyn.*, **119**, 10–19.

*Chapter 10 – UDTR rule*

LEVITT (1964).

LEVITT, H., 1971. 'Transformed up-down methods in psychoacoustics'. *J. Acoust. Soc. Amer.*, **49**, 467–477.

LEVITT, H., and BOCK, D. E., 1967. 'A sequential programmer for psychological testing'. *J. Acoust. Soc. Amer.*, **42**, 911–913.

LEVITT, H., and RABINER, L. R., 1967. Use of a sequential strategy in intelligibility testing. *J. Acoust. Soc. Amer.*, **42**, 609–612.

WETHERILL, G. B., and LEVITT, H., 'Sequential estimation of points on a psychometric function.' *Brit. J. Statist. Psych.*, **18**.

*Note*: Some further applications can be obtained from the references listed under sequential sampling in Federer and Balaam (1972).

# Author Index

# Subject Index